PROOF *of* STAKE

PROOF *of* STAKE

The Making of Ethereum, and the Philosophy of Blockchains

VITALIK BUTERIN

Edited by Nathan Schneider

SEVEN STORIES PRESS
NEW YORK · OAKLAND · LONDON

A SEVEN STORIES PRESS FIRST EDITION

Seven Stories Press
140 Watts Street
New York, NY 10013
www.sevenstories.com

Library of Congress Cataloging-in-Publication Data is on file.

ISBN: 978-1-64421-248-6 (paperback)
ISBN: 978-1-64421-249-3 (ebook)

Printed in the USA.

9 8 7 6 5 4 3 2 1

To my mom and dad,
amazing and loving parents,
entrepreneurs and internet meme lords.

TABLE OF CONTENTS

INTRODUCTION

NATHAN SCHNEIDER

Before he started building a new economic infrastructure for the internet at nineteen years old, before becoming a billionaire who sleeps on friends' couches, Vitalik Buterin wanted to write. He first became curious about Bitcoin at the urging of his father, with whom he emigrated from Russia to Canada as a child. Rather than buying, borrowing, or mining his first coins, in 2011 he posted on an online forum: Would anyone pay him with Bitcoin to write about it?

Someone did. And Buterin kept on writing, to the point of cofounding *Bitcoin Magazine*, a glossy print and digital outlet chronicling the latest developments of what was then a very small and obscure subculture. This new, hard-to-use internet money held Buterin's attention more than his first year at college did. From his time as a self-appointed reporter onward, his ideas developed in continual conversation with others. But among writings scattered over the years across various blogs, forums, and tweets, he exhibits a voice very much his own and—partly because of that voice—has built a rapt audience surrounding his invention, Ethereum. If Ethereum and its ilk become the kind of ubiquitous infrastructure they aspire to be, his ideas will need to be understood—and contested—more widely.

This book is an introduction to Vitalik Buterin the writer.

When the pseudonymous figure Satoshi Nakamoto first announced the prototype for Bitcoin in 2008, during the storm of a global financial crisis, the goal was to create a currency organized through cryptographic computer networks, rather than through governments or banks. It would come to be called a cryptocurrency. Libertarian gold-bugs and techie cypherpunks reveled in the system's metaphors: digital mining, limited supply, cash-like transactions that could be secure and private. Buterin had all the instincts of that early target audience. But as his obsession with Bitcoin deepened, by late 2013 he began to recognize that its underlying blockchain technology might be the basis of something bigger: a way of creating organizations, companies, and entire economies native to the internet. And so he wrote about it. The initial Ethereum whitepaper, included as an appendix here, lit up the still-small cryptocurrency universe when it appeared near the end of that year. Rather than depending on old-world corporations, investors, and laws to govern the servers, this would be user-governed by default. Rather than Bitcoin's metaphors of gold and mines, Ethereum culture followed the aesthetic of Buterin's favorite T-shirts, with robots, unicorns, and rainbows as the preferred mascots.

Since Ethereum went online in 2015, there have been many competing blockchains, each able to do similar things in different ways. Ethereum remains the largest among them. Although its currency, called ether or ETH, is a distant second in total value compared to Bitcoin, if you add up the value of all the products and community tokens built on top of Ethereum, it has produced the biggest share of this strange new economy. During the project's early trials, Buterin became ever more Ethereum's "benevolent dictator"—whether he liked it or not—less by any formal position than by the trust he instilled. The writings collected here have been central to building that trust.

In the process, Buterin has inhabited a space of contradiction. He wants to enable a radical reimagining of how human beings self-organize, while maintaining a rigorous agnosticism about what people choose to do with that power. "Credible neutrality," as an essay below explains, is a principle for system design, but it also describes the role he has come to play as a leader. From the earliest personnel decisions for the Ethereum Foundation to the latest high-stakes software updates, and despite his best efforts to the contrary, his leadership has been hard to distinguish from Ethereum itself. While Ethereum and systems like it are designed according to the assumption that people are selfish, he is the ascetic who seems to want nothing in particular for himself other than to enable a crypto-powered future.

There are no guarantees, however, that this will be a future worth having. When Buterin first introduced Ethereum on stage, at an early-2014 Bitcoin conference in Miami, after a litany of all the wonders that could be built with it, he ended with a mic-drop reference to Skynet—the artificial intelligence in the Terminator movies that turns on its human creators. It was a joke that he would repeat, and like many well-worn jokes it bore a warning. Ethereum holds the potential for utopia and dystopia and everything in between:

□ It creates artificial scarcity by capping the availability of made-up tokens; but these enable communities to generate abundant capital that they can use and control.

□ It excludes people who can't or won't buy and trade risky internet money; it has also spurred the invention of novel governance systems that share power with unprecedented inclusivity.

☐ It consumes vast amounts of energy just to perpetuate its own functioning; it also enables new ways of putting a price on carbon and pollution while governments refuse to do so.

☐ It has produced nouveaux riches notorious for their extravagance, congregating in tax shelters and pricing out locals; it is also a borderless, user-owned financial system available to anyone with a smartphone.

☐ It rewards a tech-savvy elite who got in early; it also presents a real chance for undermining the dominant tech companies.

☐ It has produced a speculative financial system before a real economy of useful things; yet far more than in a stock market, ownership lies with the people creating the value.

☐ It has showered vast payouts on digital collectibles with little apparent worth; the result is a new business model to support the making and sharing of open-access culture.

☐ It promises to make early adopters wealthy at the expense of future generations; it gives those generations a set of building blocks whose uses are up to those who do the building.

Readers of what follows must hold these contradictions in mind and contend with them, ascertaining for themselves and their communities which options should win out. The contradictions can be vexing and distressing, but also motivating. They are still hot enough to be shaped.

■ ■ ■

At the heart of any blockchain-based system like Bitcoin or Ethereum is the consensus mechanism. This is the process by which computers agree on a common set of data and protect it against manipulation—whether it be a list of transactions, as for Bitcoin, or the state of the Ethereum world-computer. Consensus without a central authority is not easy. Bitcoin uses a mechanism called "proof of work," which means lots of computers expend lots of energy doing math problems, all in order to prove that they are invested in keeping the system secure. The people behind those computers, known as "miners," get paid for doing so, and they consume country-sized volumes of electricity, producing the carbon emissions that level of consumption requires. Ethereum adopted proof of work as well, for want of a functional alternative at the time. But even before it went online Buterin was already talking about switching, once his team had worked out the kinks, to another mechanism: proof of stake. In proof of stake, users prove their skin in the game with token holdings rather than computing power. Energy consumption is minimal. If token holders try to corrupt the system, they lose the tokens they staked.

In this book, the consensus mechanisms are metaphors as much as system designs. They evoke the labor, commitment, conviction, and coordination that these essays depict. They also exemplify the contradictions: innovation and waste, democracy and plutocracy, vibrant community and relentless mistrust. Like the mechanisms themselves, these metaphors resist idealism, pointing to the compromises necessary for even parts of a hoped-for world to survive in the real world.

The essays included here, chosen with Buterin, present a particular side of him: the social theorist and activist, a person thinking while doing and figuring out the consequences as he goes. The largely young, male, and privileged milieu of crypto culture can often seem so far removed from the kinds of problems participants purport to be trying to solve. Buterin reflects that culture. He can be technical at times, but less here than in his other writings, many of which he intended only for fellow developers. The technical parts reward the work they take to grasp; even with formulas he is companionable, lucid, and funny.

The essays have been edited lightly for stylistic consistency. References to hyperlinks not accessible in the genre of a self-contained book have been removed. Since they were originally written for an audience with a shared subculture, here the essays include occasional editorial notes on references that might not be evident outside the crypto-sphere.

As crypto begins to break into mainstream economic life, debates have been intensifying about whether this genie needs to be put back in the bottle, if that were even possible anymore. Perhaps by reading this book, those at first invested in *whether* will find themselves turning, with Buterin, more to the ever-expanding matters of *how*. If this really is the beginning of a new social infrastructure, the political and cultural habits we develop around crypto now will have immense consequences later. As Buterin's reflections indicate, the *how* remains very much an open question.

PART I: PREMINING

Buterin reports in a January 2014 blog post that he wrote the Ethereum whitepaper "on a cold day in San Francisco in November, as a culmination of months of thought and often frustrating work." In those months, he was half-chronicler (for his* Bitcoin Magazine*) and half-builder (pitching in on several Bitcoin-related startups), making his way among New Hampshire libertarians, expats in Zürich, Tel Aviv coders, and the denizens of Calafou, a "postcapitalist colony" in a crumbling factory complex near Barcelona. Bitcoin had first been announced with a whitepaper, and crypto projects since had adopted the same form of release: even before the software, promulgate a document that is both manifesto and technical spec. The genre was well suited to Buterin's writer-builder career path in 2013. "Ethereum: A Next-Generation Cryptocurrency and Decentralized Application Platform" serves as an excellent summary for the full whitepaper, which appears in this book as an appendix. Still a year and a half before Ethereum's first release, he is already ruminating there about Ethereum 2.0 and proof of stake, which wouldn't come to fruition until 2022.*

"Premining" refers to the creation of tokens before a blockchain becomes public. By selling premined ETH on the basis of the Ethereum whitepaper, Buterin and his early collaborators raised over $18 million in bitcoin. This set a record as the largest online crowdfunding

* Vitalik Buterin, "Ethereum: Now Going Public," *Ethereum Foundation Blog*, January 23, 2014.

campaign at the time—since then exceeded mostly by projects crowdfunding on Ethereum itself. Against pressure from older, more experienced collaborators who wanted a for-profit company, Buterin insisted on building Ethereum through a nonprofit foundation. But this was not charity. He and his cofounders stood to gain significantly from the value of their premined tokens, if any of it worked.

These essays trace Buterin's evolution from a cyber-libertarian partisan to a pragmatic, big-tent infrastructure builder. At first he cheers on the Bitcoin-related projects bubbling up at the time; very few of them still exist today. The later, more chastened "On Silos" shows Buterin to be reluctant to find answers in any one project. Enabling people to radically rewrite their social contracts, Buterin argues, requires tooling that isn't loyal to any single ideology.

In the lead-up to Ethereum's release, Buterin asks himself, "Ultimately, what is it even useful for?" He outlines a theory of change based less on grand disruptions than in solving problems around the margins. The beliefs motivating the builders of this technology, he predicts, will be subsumed under what others create with it. While preparing for the public launch, his reflections increasingly center on what nobody could know or control.

—NS

MARKETS, INSTITUTIONS, AND CURRENCIES—A NEW METHOD OF SOCIAL INCENTIVIZATION

Bitcoin Magazine
January 10, 2014

Up until this point the problem of incentivizing productive activity has essentially been dominated by two major categories of solutions: markets and institutions. Markets, in their pure form, are fully decentralized, made up of a near-infinite number of agents, all engaging with each other in one-on-one interactions, each of which leaves both participants better off than they were before. Institutions, on the other hand, are inherently top-down; an institution has some governance structure that determines what the most useful activities are at any given time, and assigns a reward for people to do them. An institution's centralization allows it to incentivize production of public goods that benefit thousands or even millions of people, even if the benefit to each person is extremely small; on the other hand, as we all know, centralization has risks of its own. And for the last ten thousand years, these two options were essentially all that we had. With the rise of Bitcoin and its derivatives, however, that may all be about to change, and we may in fact now be seeing the dawn of a third form of incentivization: currencies.

THE OTHER SIDE OF THE COIN

In the standard account, a currency has three fundamental functions to society. It serves as a medium of exchange, allowing people to buy and sell goods for currency rather than having to find someone who simultaneously has exactly what you want, and wants exactly what you have, and barter with them, as a store of value, allowing people to produce and consume at different times, and as a medium of account, or a measuring stick which people can use to measure a constant "quantity of production." What many people do not realize, however, is that there is also a fourth role that currencies play, and one whose significance has been hidden throughout most of history: seigniorage.

Seigniorage can be formally defined as the difference between the market value of a currency and its intrinsic value—that is, the value that the currency would have if no one used it as currency. For ancient currencies like grain, the seigniorage was essentially zero; however, as economies and currencies got more and more complex, this "phantom value" generated by money seemingly out of nowhere would grow bigger and bigger, eventually reaching the point where, in the case of modern currencies like the dollar and the bitcoin, the seigniorage represents the entire value of the currency.

But where does the seigniorage go? In the case of currencies based on natural resources, like gold, much of the value is simply lost. Every single gram of gold comes into existence through a miner producing it; at first, some miners do earn a profit, but in an efficient market all of the easy opportunities rapidly get taken up and the cost of production approaches the return. There are of course clever ways that seigniorage can still be extracted from gold; in ancient societies, for example, kings would mint gold coins which would be worth more than ordinary gold because the coins contain an implicit promise from the king that they are not fake. In general, however, the value would not go to anyone in particular. In the case of the US dollar, we saw a slight improvement: some of the seigniorage would go to the US government. This was in many ways a large step forward, but in other ways it is also a revolution incomplete—currency, having gained the benefits of centralized seigniorage, also gained its risks by embedding itself into the heart of one of the largest centralized institutions in human history.

BITCOIN CAME ALONG

Five years ago, a new kind of money, Bitcoin, came along. In the case of Bitcoin, just like the dollar, the currency's value is 100% seigniorage; a bitcoin has no intrinsic value. But where does the seigniorage go? The answer is, some goes into the hands of the miners as profit, and the rest goes to fund the miners' expenses— the expenses of securing the Bitcoin network. Thus, in this case, we have a currency whose seigniorage goes directly into funding a public good, namely the security of the Bitcoin network itself. The importance of this is massively understated; here, we have an incentivization process that is simultaneously decentralized, requiring no authority or control, and produces a public good, all out of the ethereal "phantom value" that is somehow produced from people using Bitcoin between each other as a medium of exchange and store of value.

From there, we saw the emergence of Primecoin, the first currency that attempted to use its seigniorage for a purpose that is useful outside of itself: rather than having miners compute ultimately useless SHA256 hashes, Primecoin requires miners to find Cunningham chains of prime numbers, both supporting a very narrow category of scientific computation and providing an incentive for computer manufacturers to figure out how to better optimize circuits for arithmetic computations. And its value rapidly rose and the currency still remains the eleventh most popular today—even though its main practical benefit for each individual user, the sixty-second block time, is shared by many other currencies far more obscure than it is.

A few months later, in December, we saw the rise of a currency that is even more eccentric and surprising in its success: Dogecoin. Dogecoin, currency symbol DOGE, is a currency which, on a technical level, is almost completely identical to Litecoin; the only difference is that the maximum supply will be 100 billion instead of 84 million. But even still, the currency hit a peak market cap of over $14 million, making it the sixth largest in the world, and was even mentioned on *Business Insider* and *Vice*. So what is so special about DOGE? Essentially, the internet meme. "Doge," a slang term for "dog" that first appeared in Homestar Runner's puppet show in 2005, has since caught on as a worldwide phenomenon with the accompanying practice of putting phrases such as "wow," "so style," and "such awesome" in colorful Comic Sans font with the background of a Shiba Inu dog. This meme represents the entirety of Dogecoin's branding; all of its community websites and forums, including the official Dogecoin website, the obligatory Bitcointalk launch thread* and the /r/dogecoin and /r/dogecoinmarkets subreddits are all splattered with Doge iconography. And that's all it took to get a Litecoin clone to $14 million.

Finally, a third example comes from outside the cryptocurrency space in the form of Ven, a more traditional centralized currency which is backed by a basket of goods including commodities, currencies, and futures. Recently, Ven added carbon futures to its basket, making Ven the first currency that is in some fashion "linked to the environment." The reason this was done is a clever economic hack: the carbon futures are actually included in Ven negatively, so the value of the currency goes up as society moves away from high-CO_2 methods of production and CO_2 emissions permits become less lucrative. Thus, each individual Ven holder is now, albeit slightly, economically incentivized to support environmentally friendly living, and people are interested in Ven at least partially because of this feature.

On the whole, what these examples show is that alternative currencies are pretty much entirely dependent on grassroots marketing in order to achieve adoption; nobody takes Bitcoin or Primecoin or Dogecoin or Ven from salespeople going door to door or convincing merchants to accept them, and it is not just the technical superiority of a currency that determines its traction—ideals matter just as much. It was Bitcoin's ideals that convinced WordPress, Mega, and now Overstock to accept Bitcoin, and it is arguably for the same reason that Ripple as a payment method, despite its technical superiority to Bitcoin for merchants (specifically, five-second

---

* During this period, Bitcointalk was an online bulletin board that was a primary discussion forum for cryptocurrencies. It was founded by Satoshi Nakamoto. Any new cryptocurrency would have a forum thread associated with it.

confirmation times), has so far failed to gain much traction—its nature as a semi-centralized protocol backed by a corporation that issued 100% of the currency supply for itself makes it unattractive to many cryptocurrency enthusiasts who are interested in fairness and decentralization. And now, it is the ideals of Primecoin and Dogecoin—those of supporting science and supporting fun, respectively, that keep both of those currencies alive.

CRYPTOCOINS AS ECONOMIC DEMOCRACY

These four examples, together with this idea of phantom seigniorage value, form what is potentially a blueprint for a new kind of "economic democracy": it is possible to set up currencies whose seigniorage, or issuance, goes to support certain causes, and people can vote for those causes by accepting certain currencies at their businesses. If one does not have a business one can participate in the marketing effort and lobby other businesses to accept the currency instead. Someone can create SocialCoin, the currency which gives one thousand units per month to every person in the world, and if enough people like the idea and start accepting it, the world now has a citizen's dividend program, with no centralized funding required. We can also create currencies to incentivize medical research, space exploration, and even art; in fact, there are artists, podcasts, and musicians thinking about creating their own currencies for this exact purpose today.

In the case of one particular public good, computational research, we can actually go even further and make the distribution process happen automatically. Computational research can be incentivized by a mechanism that has not yet seen any substantial applications in the real world, but has been theorized by Peercoin and Primecoin inventor Sunny King: "proof of excellence." The idea behind proof of excellence is that the size of one's stake in the currency's decentralized voting pool and one's reward is based not on the computational power that one has or the number of coins one already owns, but on one's ability to solve complex mathematical or algorithmic challenges whose solutions would benefit all of humanity. For example, if one wants to incentivize research in number theory, one can insert the RSA integer factoring challenges into the currency, and have the currency offer fifty thousand units, plus perhaps the ability to vote on valid blocks in the mining process, automatically to the first person to provide a solution to the problem. Theoretically, this can even become a standard component in any currency's issuance model.

Of course, the idea behind using currencies in this way is not new; "social currencies" operating within local communities have existed for over a century. In recent decades, however, the social currencies movement has declined somewhat from its early twentieth-century peak, primarily because social currencies simply failed to achieve anything more than a very local reach, and because they did not benefit from the efficiencies of the banking system that more established currencies like the US dollar could attract. With cryptocurrencies, however, these objections are suddenly removed—cryptocurrencies are inherently global, and benefit from an incredibly powerful digital banking system baked right into their source code. Thus, now may be the perfect time for the social currencies movement to make a powerful, technologically-enabled comeback, and perhaps even shoot far beyond their role in the nineteenth and twentieth centuries to become a powerful, mainstream force in the world economy.

So where will we go from here? Dogecoin has already shown the public how easy it is to create your own currency; indeed, very recently the Bitcoin developer Matt Corallo has created a site, coingen.io, whose sole purpose is to allow users to quickly create their own Bitcoin or Litecoin clones with some parameter tweaks.

Even with the limited array of options that it currently has, the site has proven quite popular, with hundreds of coins created using the service despite the 0.05 BTC fee. Once Coingen allows users to add proof-of-excellence mining, an option for a portion of the issuance to go to a specific organization or fund, and more options for customized branding, we may well see thousands of cryptocurrencies being actively circulated around the internet. Will currencies fulfill their promise as a more decentralized, and democratic, way to pool together our money and support public projects and activities that help create the society we want to see? Maybe, maybe not. But with a new cryptocurrency being released almost every day, we are tantalizingly close to finding out.

ETHEREUM: A NEXT-GENERATION CRYPTOCURRENCY AND DECENTRALIZED APPLICATION PLATFORM

Bitcoin Magazine
January 23, 2014

Over the past year, there has been an increasingly large amount of discussion around so-called Bitcoin 2.0 protocols—alternative cryptographic networks that are inspired by Bitcoin, but which intend to make the underlying technology usable for far more than just currency. The earliest implementation of this idea was Namecoin, a Bitcoin-like currency created in 2010, which would be used for decentralized domain-name registration. More recently, we have seen the emergence of colored coins, allowing users to create their own currencies on the Bitcoin network, and more advanced protocols like Mastercoin, BitShares, and Counterparty, which intend to provide features such as financial derivatives, savings wallets, and decentralized exchange. However, up until this point all the protocols that have been invented have been specialized, attempting to offer detailed feature sets targeted toward specific industries or applications usually financial in nature. Now, a group of developers, including myself, have come up with a project that takes the opposite track: a cryptocurrency network that intends to be as generalized as possible, allowing anyone to create specialized applications on top for almost any purpose imaginable. The project: Ethereum.

CRYPTOCURRENCY PROTOCOLS ARE LIKE ONIONS . . .

One common design philosophy among many cryptocurrency 2.0 protocols is the idea that, just like the internet, cryptocurrency design would work best if protocols split off into different layers. Under this train of thought, Bitcoin is to be thought of as a sort of TCP/IP of the cryptocurrency ecosystem, and other next-generation protocols can be built on top of Bitcoin much like we have SMTP for email, HTTP for web pages, and XMPP for chat, all on top of TCP as a common underlying data layer.

So far, the three main protocols that have followed this model are colored coins, Mastercoin, and Counterparty. The way the colored coins protocol works is simple. First, in order to create colored coins, a user tags specific bitcoins as having a special meaning; for example, if Bob is a gold issuer, he may wish to tag some set of bitcoins and say that each satoshi represents 0.1 grams of gold redeemable from him. The protocol then tracks those bitcoins through the blockchain, and in that way it is possible to calculate who owns them at any time.

Mastercoin and Counterparty are somewhat more abstract; they use the Bitcoin blockchain to store data, so a Mastercoin or Counterparty transaction is a Bitcoin transaction, but the protocols interpret the transactions in a completely different way. One can have two Mastercoin transactions, one sending 1 MSC and the other 100,000 MSC, but from the point of view of a Bitcoin user that does not know how that Mastercoin protocol works, they both look like small transactions sending 0.0006 BTC each;

the Mastercoin-specific metadata is encoded in the transaction outputs. A Mastercoin client then needs to search the Bitcoin blockchain for Mastercoin transactions in order to determine the current Mastercoin balance sheet.

I personally have had the privilege of talking directly to many of the originators of the colored coins and Mastercoin protocol, and have participated considerably in the development of both projects. However, over about two months of research and participation, what I eventually came to realize is that, while the underlying idea of having such high-level protocols on top of low-level protocols is laudable, there are fundamental flaws in the implementations, as they stand today, that may well prevent the projects from ever gaining anything more than a small amount of traction.

The reason is not that the ideas behind the protocols themselves are bad; the ideas are excellent, and the response of the community alone is proof that they are trying to do something that is very much needed. Rather, the reason is that the low-level protocol that they are trying to build their high-level protocols on top of, Bitcoin, is simply not cut out for the task. This is not to say that Bitcoin is bad, or is not a revolutionary invention; as a protocol for storing and transferring value, Bitcoin is excellent. However, as far as being an effective low-level protocol is concerned, Bitcoin is less effective; rather than being like a TCP on top of which one can build HTTP, Bitcoin is like SMTP: a protocol that is good at its intended task (in SMTP's case email, in Bitcoin's case money), but not particularly good as a foundation for anything else.

The specific failure of Bitcoin is particularly concentrated in one place: scalability. Bitcoin itself is as scalable as a cryptocurrency can be; even if the blockchain balloons to over a terabyte, there is a protocol called "simplified payment verification," described in the Bitcoin whitepaper, that allows "light clients" with only a few megabytes of bandwidth and storage to securely determine whether or not they have received transactions. With colored coins and Mastercoin, however, this possibility disappears. The reason is this. In order to determine what color a colored coin is, you need to not just use Bitcoin simplified payment verification to prove that it exists; you also need to trace it all the way back to its genesis, and do an SPV check each step of the way. Sometimes, the backward scan is exponential; and with metacoin protocols there is no way to know anything at all without verifying every single transaction.

And this is what Ethereum intends to fix. Ethereum does not intend to be a Swiss Army knife protocol with hundreds of features to suit every need; instead, Ethereum aims to be a superior foundational protocol, and allow other decentralized applications to build on top of it instead of Bitcoin, giving them more tools to work with and allowing them to gain the full benefits of Ethereum's scalability and efficiency.

CONTACTS, NOT JUST FOR DIFFERENCE

At the time that Ethereum was being developed, there was a large amount of interest in allowing financial contracts on top of cryptocurrencies; the basic type of contract being a "contract for difference." In a contract for difference, two parties agree to put in some amount of money, and then get money out in a proportion that depends on the value of some underlying asset. For example, a CFD might have Alice put in $1,000, Bob put in $1,000, and then after thirty days the blockchain would automatically return to Alice $1,000 plus $100 for every dollar that the LTC/USD price went up during that time period, and send Bob the rest. These contracts allow people to speculate on assets at high leverage, or alternatively protect themselves from cryptocurrency volatility by canceling out their exposure, without any centralized exchange.

At this point, however, it is clear that contracts for difference are really only one special case of a much more general concept: contracts for formulas. Instead of having the contract take in $x from Alice, $y from Bob, and return to Alice $x plus an additional $z for every dollar that some given ticker went up, a contract should be able to return to Alice an amount of funds based on any mathematical formula, allowing contracts of arbitrary complexity. If the formula allows random data as inputs, these generalized CFDs can even be used to implement a sort of peer-to-peer gambling.

Ethereum takes this idea and pushes it one step further. Instead of contracts being agreements between two parties that start and end, contracts in Ethereum are like a sort of autonomous agent simulated by the blockchain. Each Ethereum contract has its own internal scripting code, and the scripting code is activated every time a transaction is sent to it. The scripting language has access to the transaction's value, sender, and optional data fields, as well as some block data and its own internal memory, as inputs, and can send transactions. To make a CFD, Alice would create a contract and seed it with $1,000 worth of cryptocurrency, and then wait for Bob to accept the contract by sending a transaction containing $1,000 as well. The contract would then be programmed to start a timer, and after thirty days Alice or Bob would be able to send a small transaction to the contract to activate it again and release the funds.

Code example of an Ethereum currency contract, written in a high-level language:

```
if tx.value < 100 * block.basefee:
  stop
```

```
if contract.memory[1000]:
  from = tx.sender
  to = tx.data[0]
  value = tx.data[1]
  if to <= 1000:
    stop
  if contract.memory[from] < value:
    stop
  contract.memory[from] = contract.memory[from] - value
  contract.memory[to] = contract.memory[to] + value
else: contract.memory[mycreator] = 10000000000000000
contract.memory[1000] = 1
```

Aside from this narrow contract-for-difference model, however, the whitepaper outlines many other transaction types that will become possible with Ethereum scripting, of which a few include:

□ **MULTISIGNATURE ESCROWS:** Of a similar spirit to the Bitcoin arbitration service Bitrated, but with more complex rules than Bitcoin. For example, there will be no need for the signers to pass around partially signed transactions manually; people can authorize a withdrawal asynchronously over the blockchain one at a time and then have the transaction finalized automatically once enough people make their authorizations.

□ **SAVINGS ACCOUNTS:** One interesting setup works as follows. Suppose that Alice wants to store a large amount of money, but does not want to risk losing everything if her private key is lost or stolen. She makes a contract with Bob, a semi-trustworthy bank, with the following rules: Alice is allowed to withdraw up to 1 per day, Alice with

Bob's approval can withdraw any amount, and Bob alone can withdraw up to 0.05 per day. Normally, Alice will only need small amounts at a time, and if Alice wants more she can prove her identity to Bob and make the withdrawal. If Alice's private key gets stolen, she can run to Bob and move the funds into another contract before the thief gets away with more than 1 of the funds. If Alice loses her private key, Bob will eventually be able to recover her funds. And if Bob turns out to be evil, Alice can withdraw her own funds twenty times faster than he can. In short, all of the security of traditional banking, but with almost none of the trust.

☐ **PEER-TO-PEER GAMBLING:** Any kind of peer-to-peer gambling protocol can be implemented on top of Ethereum. A very basic protocol would simply be a contract for difference on random data such as a block hash.

☐ **CREATING YOUR OWN CURRENCY:** Using Ethereum's internal memory store, you can create an entire new currency inside of Ethereum. These new currencies can be constructed to interact with each other, have a decentralized exchange, or any other kind of advanced features.

This is the advantage of Ethereum code: because the scripting language is designed to have no restrictions except for a fee system, essentially any kind of rules can be encoded inside of it. One can even have an entire company manage its savings on the blockchain, with a contract saying that, for example, sixty of the current shareholders of a company are needed to agree to move any funds (and perhaps thirty can move a maximum of 1 per day). Other, less traditionally capitalistic structures are also possible; one idea is for a democratic organization with the only rule being that two-thirds of the existing members of a group must agree to invite another member.

BEYOND THE FINANCIAL

The financial applications, however, only scratch the surface of what Ethereum, and cryptographic protocols on top of Ethereum, can do. While Ethereum's financial applications may be what initially excites many people in the cryptocurrency community, the long-term promise is arguably in the ways that Ethereum can work together with other, non-financial peer-to-peer protocols. One of the main problems that non-financial peer-to-peer protocols have faced so far is the lack of incentive—that is to say, unlike centralized for-profit platforms, there is no financial reason to participate. In some cases, participation is in some sense its own reward; it is for this reason that people continue to write open-source software, contribute to Wikipedia, and make comments on forums and write blog posts. In the context of peer-to-peer protocols, however, participation is often not a "fun" activity in any meaningful sense; rather, it consists of putting in a large quantity of resources, letting a daemon run in the background (potentially hogging CPU and battery power), and forgetting about it.

For example, there have already for a long time been data protocols, such as Freenet, that essentially provide everyone with decentralized uncensorable static-content hosting; in practice, however, Freenet is very slow, and few people contribute resources. File-sharing protocols all suffer from the same problem: although altruism is good enough for spreading popular commercial block-busters around, it becomes markedly less effective for those with less mainstream preferences. Thus, perversely, the peer-to-peer nature of file sharing may actually be helping the centralization of entertainment and media production, not hindering it. All of these problems, however, can potentially be solved if we add incentivization—empowering people to build not just nonprofit side projects but also businesses and livelihoods around participating in the network.

□ **INCENTIVIZED DATA STORAGE:** Essentially, a decentralized Dropbox. The idea works as follows: if a user wants to have a 1 GB file backed up by the network, they would construct a data structure known as a Merkle tree out of the data. They would then put the root of the tree, along with 10 ether, into a contract and upload the file onto another specialized network that nodes wishing to rent out their hard-drive space would listen for messages on. Every day, the contract would automatically pick a random branch of the tree (e.g., "left → right → left → left → left → right → left"), ending at a block of the file, and giving 0.01 ether to the first node to provide that branch. Nodes would store the entire file to maximize their chance of getting the reward.

□ **BITMESSAGE AND TOR:** Bitmessage is a next-generation email protocol that is both fully decentralized and encrypted, allowing anyone to send messages to any other Bitmessage user securely without relying on any third parties except for the network. However, Bitmessage has one large usability flaw: instead of sending messages to human-friendly email addresses, like "myname@email," you need to send to garbled thirty-four-character Bitmessage addresses (e.g., "BM-BcbRqcFFSQUUmXFKsPJgVQPSiFA3Xash"). Ethereum contracts offer a solution: people can register their names on a special Ethereum contract, and Bitmessage clients can query the Ethereum blockchain to get the thirty-four-character Bitmessage address associated with any name behind the scenes. The online anonymizing network Tor suffers from the same problems, and thus can also benefit from this solution.

◻ **IDENTITY AND REPUTATION SYSTEMS:** Once you can register your name on the blockchain, the logical next step is obvious: have a web of trust on the blockchain. Webs of trust are a key part of an effective peer-to-peer communication infrastructure: you don't just want to know that a given public key refers to a given person; you also want to know that the person is trustworthy in the first place. The solution is to use social networks: if you trust A, A trusts B, and B trusts C, then there is a pretty good chance that you can trust C, at least to some extent. Ethereum can serve as the data layer for a fully decentralized reputation system—and potentially ultimately a fully decentralized marketplace.

Many of the above applications consist of actual peer-to-peer protocols and projects that are already well under development; in those cases, we intend to establish partnerships with as many of these projects as we can, and help fund them in exchange for bringing their value into the Ethereum ecosystem. We want to help not just the cryptocurrency community, but also the peer-to-peer community as a whole, including file sharing, torrents, data storage, and mesh networking. We believe that there are many projects, especially in the non-financial area, that can potentially bring great value to the community, but for which development is underfunded precisely because they lack an opportunity to effectively introduce a financial component; perhaps Ethereum may be what ultimately pushes dozens of these projects to the next stage.

Why are all of these applications possible on top of Ethereum? The answer lies in the currency's internal programming language. An analogy here may be made with the internet. Back in 1996, the web was nothing but HTML, and all people could do with it was serve static web pages on sites like GeoCities. Then developers decided that there was a great need for people to submit forms in HTML, so HTML added a forms feature. This was like a "colored coins" of web protocols: try to solve a specific problem, but do it on top of a weak protocol without looking at the larger picture. Soon, however, we came up with JavaScript, a programming language inside the web browser. And it was JavaScript that solved the problem: because JavaScript is a universal, Turing-complete programming language, it can be used to build apps of arbitrary complexity; Gmail, Facebook, and even Bitcoin wallets have all been made with the language. And this was not because the JavaScript developers decided that they wanted people to build Gmail, Facebook, and Bitcoin wallets; they just wanted a programming language. What we can do with the language is up to our own imaginations. And that is the spirit that we want to bring to Ethereum. Ethereum does not intend to be the end of all cryptocurrency innovation; it intends to be the beginning.

FURTHER INNOVATIONS

Along with its main feature of a Turing-complete, universal scripting language, Ethereum will also have a number of other improvements over existing cryptocurrency:

☐ **FEES:** Ethereum contracts will regulate its Turing-complete functionality and prevent abusive transactions such as memory hogs and infinite-loop scripts by instituting a transaction fee for each computational step of script execution. More expensive operations, such as storage accesses and cryptographic operations, will have higher fees, and there will also be a fee for every item of storage that a contract fills up. To encourage contracts to clean up after themselves, if a contract reduces the amount of storage that it uses, a negative fee will be charged; in fact, there is a special SUICIDE opcode to clear a contract and send all funds and the hefty negative fee back to its creator.

☐ **MINING ALGORITHMS:** There has been a lot of interest in making cryptocurrencies whose mining is resistant against specialized hardware, allowing ordinary users with commodity hardware to participate without any capital investment and helping to avoid centralization. So far, the main antidote has been Scrypt, a mining algorithm that requires a large amount of both computational power and memory; however, Scrypt is not memory-hard enough, and there are companies building specialized devices for it. We have come up with Dagger, a prototype proof of work that is even more memory-hard than Scrypt, as well as prototype proof-of-stake algorithms such as Slasher that get around the issue of mining entirely. Ultimately, however, we intend to host a contest, similar to the contests that determined the standards for AES and SHA3, where we invite research groups from universities around the world to devise the best possible commodity-hardware-friendly mining algorithm.

☐ **GHOST:** GHOST is a new block propagation protocol pioneered by Aviv Zohar and Yonatan Sompolinsky that allows blockchains to have much faster block confirmation times, ideally in the range of three to thirty seconds, without running into the issues of centralization and high stale rate that fast block confirmations normally bring. Ethereum is the first major currency to integrate a simplified single-level version of GHOST as part of its protocol.

THE PLAN

Ethereum is potentially a massive and wide-reaching undertaking, and will take months to develop. With that in mind, the currency will be released in multiple stages. The first stage, the release of the whitepaper, has already happened. Forums, a wiki, and a blog have been set up, and anyone is free to visit them and set up an account and comment on the forums. On January 25, a sixty-day fundraiser will launch at the conference in Miami, during which anyone will be able to purchase ether, Ethereum's internal currency, for BTC, much like the Mastercoin fundraiser; the price will be 1,000 ETH for 1 BTC, although early investors will get roughly a $2x$ benefit to compensate for the increased risk that they're taking for participating in the project earlier. The fundraiser participants will not just get ether; there will also be a number of additional rewards, likely including free tickets to conferences, a spot to put thirty-two bytes into the genesis block, and for the top donors, even the ability to name three subunits of the currency (e.g., the equivalent of the "microbitcoin" in BTC).

The issuance of Ethereum will not be any single mechanism; instead, a compromise approach combining the benefits of multiple approaches will be used. The issuance model will work as follows:

Ether will be released in a fundraiser at the price of 1,000 to 2,000 ETH per BTC, with earlier funders getting a better price to compensate for the increased uncertainty of participating at an earlier stage. The minimum funding amount will be 0.01 BTC. Suppose that x ETH gets released in this way:

☐ 0.225x ETH will be allocated to the fiduciary members and early contributors who substantially participated in the project before the start of the fundraiser. This share will be stored in a timelock contract; about 40% of it will be spendable after one year, 70% after two years, and 100% after three years.

☐ 0.05x ETH will be allocated to a fund to use to pay expenses and rewards in ether between the start of the fundraiser and the launch of the currency.

☐ 0.225x ETH will be allocated as a long-term reserve pool to pay expenses, salaries, and rewards in ether after the launch of the currency.

☐ 0.4x ETH will be mined per year forever after that point.

There is an important distinction compared to Bitcoin and most other cryptocurrencies: here, the eventual supply is unlimited. The "permanent linear inflation" model is designed to make ether neither inflationary or deflationary; the lack of a supply cap is intended to dampen some of the speculative and wealth-inequality effects of existing currencies, but at the same time the linear, rather than traditionally exponential, inflation model will mean that the effective inflation rate tends to zero over time. Additionally, because the initial currency supply will not start from zero, the currency supply growth in the first eight years will actually be slower than Bitcoin, giving fundraiser participants and early adopters a chance to benefit substantially in the medium term.

At some point in February, we will release a centralized testnet—a server which anyone can use to send transactions and create contracts. Soon after that, the decentralized testnet will come, which we will use to test different mining algorithms and make sure that the peer-to-peer daemon works and is secure, and take measurements to look for optimizations to the scripting language. Finally, once we are sure that the protocol and the client is secure, we will release the genesis block and allow mining to begin.

LOOKING FORWARD

Since Ethereum includes a Turing-complete scripting language, it can be mathematically proven that it can do essentially anything that a Bitcoin-like blockchain-based cryptocurrency potentially can do. But there are still problems that the protocol, as it stands today, leaves unresolved. For example, Ethereum offers no solution for the fundamental scalability problem in all blockchain-based cryptocurrencies—namely, the fact that every full node must store the entire balance sheet and verify every transaction. Ethereum's concept of a separate "state tree" and "transaction list," borrowed from Ripple, mitigates this to some extent, but nevertheless no fundamental breakthrough is mine. For that, technology like Eli Ben-Sasson's Secure Computational Integrity and Privacy (SCIP), now under development, will be required.

Additionally, Ethereum offers no improvements on traditional proof-of-work mining with all its flaws, and proof-of-excellence and Ripple-style consensus are left unexplored. If it turns out that proof-of-stake or some other proof-of-work algorithm is a better solution, then future cryptocurrencies may use proof-of-stake algorithms like MC2 and Slasher instead. If there is room for an Ethereum 2.0, it is in these areas that the improvements will lie. And ultimately, Ethereum is an open-ended project; if the project gets enough funding, we may even be the ones to release Ethereum 2.0 ourselves, carrying over the original account balances onto an even further improved network. Ultimately, as in our slogan for the currency itself, the only limit is our imagination.

SELF-ENFORCING CONTRACTS AND FACTUM LAW

Ethereum blog
February 24, 2014

Many of the concepts that we promote over in Ethereum land may seem incredibly futuristic, and perhaps even frightening, at times. We talk about so-called "smart contracts" that execute themselves without any need, or any opportunity, for human intervention or involvement, people forming Skynet-like "decentralized autonomous organizations" that live entirely on the cloud and yet control powerful financial resources and can incentivize people to do very real things in the physical world, decentralized "math-based law," and a seemingly utopian quest to create some kind of fully trust-free society. To the uninformed user, and especially to those who have not even heard of plain old Bitcoin, it can be hard to see how these kinds of things are possible and, if they are, why they can possibly be desirable. The purpose of this series will be to dissect these ideas in detail, and show exactly what we mean by each one, discussing its properties, advantages, and limitations.

The first installment of the series will talk about so-called smart contracts. Smart contracts are an idea that has been around for several decades, but was given its current name and first substantially brought to the (cryptography-inclined) public's attention by Nick Szabo in 2005. In essence, the definition of a smart contract is simple: a smart contract is a contract that enforces itself. That is to say, whereas a regular contract is a piece of paper (or, more recently, a PDF document) containing text which implicitly asks for a judge to order a party to send money (or other property) to another party under certain conditions, a smart contract is a computer program that can be run on hardware which automatically executes those conditions. Nick Szabo uses the example of a vending machine:

> A canonical real-life example, which we might consider to be the primitive ancestor of smart contracts, is the humble vending machine. Within a limited amount of potential loss (the amount in the till should be less than the cost of breaching the mechanism), the machine takes in coins, and via a simple mechanism, which makes a freshman computer science problem in design with finite automata, dispense[s] change and product according to the displayed price. The vending machine is a contract with the bearer: anybody with coins can participate in an exchange with the vendor. The lockbox and other security mechanisms protect the stored coins and contents from attackers, sufficiently to allow profitable deployment of vending machines in a wide variety of areas.

Smart contracts are the application of this concept to, well, lots of things. We can have smart financial contracts that automatically shuffle money around based on certain formulas and conditions, smart domain-name sale orders that give the domain to whoever first sends in two hundred dollars, perhaps even smart insurance contracts that control bank accounts and automatically pay out based on some trusted source (or combination of sources) supplying data about real-world events.

SMART PROPERTY

At this point, however, one obvious question arises: How are these contracts going to be enforced? Just like traditional contracts, which are not worth the paper they're written on unless there's an actual judge backed by legal power enforcing them, smart contracts need to be "plugged in" to some system in order to actually have power to do anything. The most obvious, and oldest, solution is hardware, an idea that also goes by the name "smart property." Nick Szabo's vending machine is the canonical example here. Inside the vending machine, there is a sort of proto-smart-contract, containing a set of computer code that looks something like this:

```
if button_pressed == "Coca-Cola" and money_inserted
>= 1.75:
   release("Coca-Cola")
   return_change(money_inserted - 1.75)
else if button_pressed == "Aquafina Water" and
money_inserted
>= 1.25:
   release("Aquafina Water")
   return_change(money_inserted - 1.25)
else if …
```

The contract has four "hooks" into the outside world: the button pressed and money inserted variables as input, and the release and return change commands as output. All four of these depend on hardware, although we focus on the last three because human input is generally considered to be a trivial problem. If the contract was running on an Android phone from 2007, it would be useless; the Android phone has no way of knowing how much money was inserted into a slot, and certainly cannot release Coca-Cola bottles or return change. On a vending machine, on the other hand, the contract carries some "force," backed by the vending machine's internal Coca-Cola holdings and its physical security preventing people from just taking the Coca-Cola without following the rules of the contract.

Another, more futuristic, application of smart property is rental cars: imagine a world where everyone has their own private key on a smartphone, and there is a car such that when you pay one hundred dollars to a certain address the car automatically starts responding commands signed by your private key for a day. The same principle can also be applied to houses. If that sounds farfetched, keep in mind that office buildings are largely smart property already: access is controlled by access cards, and the question of which (if any) doors each card is valid for is determined by a piece of code linked to a database. And if the company has an HR system that automatically processes employment contracts and activates new employees' access cards, then that employment contract is, to a slight extent, a smart contract.

SMART MONEY AND FACTUM SOCIETY

However, physical property is very limited in what it can do. Physical property has a limited amount of security, so you cannot practically do anything interesting with more than a few tens of thousands of dollars with a smart-property setup. And ultimately, the most interesting contracts involve transferring money. But how can we actually make that work? Right now, we basically can't. We can, theoretically, give contracts the login details to our bank accounts, and then have the contract send money under some conditions, but the problem is that this kind of contract is not really "self-enforcing." The party making the contract can always simply turn the contract off just before payment is due, or drain their bank account, or even simply change the password to the account. Ultimately, no matter how the contract is integrated into the system, someone has the ability to shut it off.

How can we solve the problem? Ultimately, the answer is one that is radical in the context of our wider society, but already very much old news in the world of Bitcoin: we need a new kind of money. So far, the evolution of money has followed three stages: commodity money, commodity-backed money, and fiat money. Commodity money is simple: it's money that is valuable because it is also simultaneously a commodity that has some "intrinsic" use value. Silver and gold are perfect examples, and in more traditional societies we also have tea, salt (etymology note: this is where the word "salary" comes from), seashells, and the like. Next came commodity-backed money—banks issuing certificates that are valuable because they are redeemable for gold. Finally, we have fiat money. The "fiat" in "fiat money" is just like in "fiat lux," except instead of God saying "let there be light" it's the federal government saying "let there be money." The money has value largely because the government issuing it accepts that money, and only that money, as payment for taxes and fees, alongside several other legal privileges.

With Bitcoin, however, we have a new kind of money: factum money. The difference between fiat money and factum money is this: whereas fiat money is put into existence, and maintained, by a government (or, theoretically, some other kind of agency) producing it, factum money just is. Factum money is simply a balance sheet, with a few rules on how that balance sheet can be updated, and that money is valid among that set of users which decides to accept it. Bitcoin is the first example, but there are more. For example, one can have an alternative rule, which states that only bitcoins coming out of a certain "genesis transaction" count as part of the balance sheet; this is called "colored coins," and is also a kind of factum money (unless those colored coins are fiat or commodity-backed).

The main promise of factum money, in fact, is precisely the fact that it meshes so well with smart contracts. The main problem with smart contracts is enforcement: if a contract says to send $200 to Bob if X happens, and X does happen, how do we ensure that $200 actually gets sent to Bob? The solution with factum money is incredibly elegant: *the definition of the money*, or more precisely, the definition of the current balance sheet, is the result of executing all of the contracts. Thus, if X does happen, then everyone will agree that Bob has the extra $200, and if X does not happen, then everyone will agree that Bob has whatever Bob had before.

This is actually a much more revolutionary development than you might think at first; with factum money, we have created a way for contracts, and perhaps even law in general, to work, and be effective, without relying on any kind of mechanism whatsoever to enforce it. Want a hundred dollar fine for littering? Then define a currency so that you have one hundred units less if you litter, and convince people to accept it. Now, that particular example is very farfetched, and likely impractical without a few major caveats, which we will discuss below, but it shows the general principle, and there are many more moderate examples of this kind of principle that definitely can be put to work.

JUST HOW SMART ARE SMART CONTRACTS?

Smart contracts are obviously very effective for any kind of financial application, or more generally any kind of swap between two different factum assets. One example is a domain-name sale; a domain, like google.com, is a factum asset, since it's backed by a database on a server that only carries any weight because we accept it, and money can obviously be factum as well. Right now, selling a domain is a complicated process that often requires specialized services; in the future, you may be able to package up a sale offer into a smart contract and put it on the blockchain, and if anyone takes it both sides of the trade will happen automatically—no possibility of fraud involved. Going back to the world of currencies, decentralized exchange is another example, and we can also do financial contracts such as hedging and leverage trading.

However, there are places where smart contracts are not so good. Consider, for example, the case of an employment contract: A agrees to do a certain task for B in exchange for payment of x units of currency C. The payment part is easy to smart-contract-ify. However, there is a part that is not so easy: verifying that the work actually took place. If the work is in the physical world, this is pretty much impossible, since blockchains don't have any way of accessing the physical world. Even if it's a website, there is still the question of assessing quality, and although computer programs can use machine-learning algorithms to judge such characteristics quite effectively in certain cases, it is incredibly hard to do so in a public contract without opening the door for employees "gaming the system." Sometimes, a society ruled by algorithms is just not quite good enough.

Fortunately, there is a moderate solution that can capture the best of both worlds: judges. A judge in a regular court has essentially unlimited power to do what they want, and the process of judging does not have a particularly good interface; people need to file a suit, wait a significant length of time for a trial, and the judge eventually makes a decision which is enforced by the legal system—itself not a paragon of lightning-quick efficiency. Private arbitration often manages to be cheaper and faster than courts, but even there the problems are still the same. Judges in a factum world, on the other hand, are very much different. A smart contract for employment might look like this:

```
if says(B,"A did the job") or says(J,"A did the
job"):
  send(200, A)
else if says(A,"A did not do the job") or says(J,"A
did not do the job"):
  send(200, B)
```

says is a signature-verification algorithm; says(P,T) basically checks if someone had submitted a message with text T and a digital signature that verifies using P's public key. So how does this contract work? First, the employer would send 200 currency units into the contract, where they would sit in escrow. In most cases, the employer and employee are honest, so either A quits and releases the funds back to B by signing a message saying "A did not do the job," or A does the job, B verifies that A did the job, and the contract releases the funds to A. However, if A does the job, and B disagrees, then it's up to judge J to say that either A did the job or A did not do the job.

Note that J's power is very carefully delineated; all that J has the right to do is say that either A did the job or A did not do the job. A more sophisticated contract might also give J the right to grant judgments within the range between the two extremes. J does not have the right to say that A actually deserves 600 currency units, or that by the way the entire relationship is illegal and J should get the 200 units, or anything else outside of the clearly defined boundaries. And J's power is enforced by factum—the contract contains J's public key, and thus the funds automatically go to A or B based on the boundaries. The contract can even require messages from two out of three judges, or it can have separate judges judge separate aspects of the work and have the contract automatically assign B's work a quality score based on those ratings. Any contract can simply plug in any judge in exactly the way that they want, whether to judge the truth or falsehood of a specific fact, provide a measurement of some variable, or be one of the parties facilitating the arrangement.

How will this be better than the current system? In short, what this introduces is "judges as a service." Now, in order to become a "judge" you need to get hired at a private arbitration firm or a government court or start your own. In a cryptographically enabled factum law system, being a judge simply requires having a public key and a computer with internet access. As counterintuitive as it sounds, not all judges need to be well-versed in law. Some judges can specialize in, for example, determining whether or not a product was shipped correctly (ideally, the postal system would do this). Other judges can verify the completion of employment contracts. Others would appraise damages for insurance contracts. It would be up to the contract writer to plug in judges of each type in the appropriate places in the contract, and the part of the contract that can be defined purely in computer code will be.

And that's all there is to it.

ON SILOS

Ethereum blog
December 31, 2014

One of the criticisms that many people have made about the current direction of the cryptocurrency space is the increasing amount of fragmentation that we are seeing. What was earlier perhaps a more tightly bound community centered around developing the common infrastructure of Bitcoin is now increasingly a collection of "silos," discrete projects all working on their own separate things. There are a number of developers and researchers who are either working for Ethereum or working on ideas as volunteers and happen to spend lots of time interacting with the Ethereum community, and this set of people has coalesced into a group dedicated to building out our particular vision. Another quasi-decentralized collective, BitShares, has set their hearts on their own vision, combining their particular combination of DPoS,* market-pegged assets, and vision of blockchain as decentralized autonomous corporation as a way of reaching their political goals of free-market libertarianism and a contract-free society. Blockstream,

---

* DPoS stands for "delegated proof of stake," a consensus mechanism that limits who can serve as validators.

51

the company behind "side chains," has likewise attracted their own group of people and their own set of visions and agendas—and likewise for Truthcoin, MaidSafe, NXT, and many others.

One argument, often raised by Bitcoin maximalists and side-chains proponents, is that this fragmentation is harmful to the cryptocurrency ecosystem—instead of all going our own separate ways and competing for users, we should all be working together and cooperating under Bitcoin's common banner. As Fabian Brian Crain summarizes:

> One recent event that has further inflamed the discussion is the publication of the side chain proposal. The idea of sidechains is to allow the trustless innovation of altcoins while offering them the same monetary base, liquidity and mining power of the Bitcoin network.
>
> For the proponents, this represents a crucial effort to rally the cryptocurrency ecosystem behind its most successful project and to build on the infrastructure and ecosystem already in place, instead of dispersing efforts in a hundred different directions.

Even to those who disagree with Bitcoin maximalism, this seems like a rather reasonable point, and even if the cryptocurrency community should not all stand together under the banner of "Bitcoin" one may argue that we need to all stand together somehow, working to build a more unified ecosystem. If Bitcoin is not powerful enough to be a viable backbone for life, the crypto universe and everything, then why not build a better and more scalable decentralized computer instead and build everything on that? Hypercubes certainly *seem* powerful enough to be worth being a maximalist over, if you're the sort of person to whom one-X-to-rule-them-all proposals are intuitively appealing, and the members of BitShares, Blockstream, and other "silos" are often quite eager to believe the same thing about their own particular solutions, whether they are based on merged-mining, DPoS plus BitAssets, or whatever else.

So why not? If there truly is one consensus mechanism that is best, why should we not have a large merger between the various projects, come up with the best kind of decentralized computer to push forward as a basis for the crypto-economy, and move forward together under one unified system? In some respects, this seems noble; "fragmentation" certainly has undesirable properties, and it is natural to see "working together" as a good thing. In reality, however, while more cooperation is certainly useful, and this blog post will later describe how and why, desires for extreme consolidation or winner-take-all are to a large degree *exactly wrong*—not only is fragmentation not all that bad, but rather it's inevitable, and arguably the only way that this space can reasonably prosper.

AGREE TO DISAGREE

Why has fragmentation been happening, and why should we continue to let it happen? To the first question, and also simultaneously to the second, the answer is simple: we fragment because we disagree. Particularly, consider some of the following claims, all of which I believe in, but which are in many cases a substantial departure from the philosophies of many other people and projects:

□   I do not think that weak subjectivity* is all that much of a problem. However, much higher degrees of subjectivity and intrinsic reliance on extra-protocol social consensus I am still not comfortable with.

---

* Weak subjectivity is a concept of Buterin's that deals with what a network node needs to know in a proof-of-stake system.

☐ I consider Bitcoin's $600 million/year wasted electricity on proof of work to be an utter environmental and economic tragedy.

☐ I believe ASICs* are a serious problem, and that as a result of them Bitcoin has become qualitatively less secure over the past two years.

☐ I consider Bitcoin (or any other fixed-supply currency) to be too incorrigibly volatile to ever be a stable unit of account, and believe that the best route to cryptocurrency price stability is by experimenting with intelligently designed flexible monetary policies (i.e., NOT "the market" or "the Bitcoin central bank"). However, I am not interested in bringing cryptocurrency monetary policy under any kind of centralized control.

☐ I have a substantially more anti-institutional/libertarian/anarchist mindset than some people, but substantially less so than others (and am incidentally not an Austrian economist). In general, I believe there is value to both sides of the fence, and believe strongly in being diplomatic and working together to make the world a better place.

☐ I am not in favor of there being one-currency-to-rule-them-all, in the crypto-economy or anywhere.

☐ I think token sales are an awesome tool for decentralized protocol monetization, and that everyone attacking the concept outright is doing a disservice to society by threatening to take away a beautiful thing. However, I do agree

---

* ASIC stands for application-specific integrated circuit. In the blockchain context, it refers to computers designed specifically for efficient "mining" in proof-of-work systems. Crypto mining centers can be warehouses full of these machines designed and built entirely to churn the otherwise useless math required to confirm blocks.

that the model as implemented by us and other groups so far has its flaws and we should be actively experimenting with different models that try to align incentives better.

□ I believe futarchy* is promising enough to be worth trying, particularly in a blockchain-governance context.

□ I consider economics and game theory to be a key part of cryptoeconomic protocol analysis, and consider the primary academic deficit of the cryptocurrency community to be not ignorance of advanced computer science, but rather of economics and philosophy. We should reach out to lesswrong.com** more.

□ I see one of the primary reasons why people will adopt decentralized technologies (blockchains, whisper, DHTs) in practice to be the simple fact that software developers are lazy, and do not wish to deal with the complexities of maintaining a centralized website.

□ I consider the blockchain-as-decentralized-autonomous-corporation metaphor to be useful, but limited. Particularly, I believe that we as cryptocurrency developers should be taking advantage of this perhaps brief period in which cryptocurrency is still an idealist-controlled industry to design institutions that maximize utilitarian social-welfare metrics, not profit (no, they are not equivalent).

There are probably very few people who agree with me on every single one of the items above. And it is not just myself that has my

---

* Futarchy is a governance model in which voters choose certain social goals, and in prediction markets, investors bet on the policies they believe are most likely to achieve those goals.
** A rationalist online community blog founded by the artificial-intelligence researcher Eliezer Yudkowsky.

own peculiar opinions. As another example, consider the fact that the CTO of Open Transactions, Chris Odom, says things like this:

> What is needed is to replace trusted entities with systems of cryptographic proof. Any entity that you see in the Bitcoin community that you have to trust is going to go away, it's going to cease to exist . . . Satoshi's dream was to eliminate [trusted] entities entirely, either eliminate the risk entirely or distribute the risk in a way that it's practically eliminated.

Meanwhile, certain others feel the need to say things like this:

> Put differently, commercially viable reduced-trust networks do not need to protect the world from platform operators. They will need to protect platform operators from the world for the benefit of the platform's users.

Of course, if you see the primary benefit of cryptocurrency as being regulation avoidance then that second quote also makes sense, but in a way completely different from the way its original author intended—but that once again only serves to show just how differently people think. Some people see cryptocurrency as a capitalist revolution, others see it as an egalitarian revolution, and others see everything in between. Some see human consensus as a very fragile and corruptible thing and cryptocurrency as a beacon of light that can replace it with hard math; others see cryptocurrency consensus as being only an extension of human consensus, made more efficient with technology. Some consider the best way to achieve crypto assets with dollar parity to be dual-coin financial derivative schemes; others see the simpler approach as being to use blockchains to represent claims on real-world assets instead (and still others think that Bitcoin will eventually be more stable than the dollar all on its own). Some think that scalability is best done by "scaling up"; others believe the ultimately superior option is "scaling out."

Of course, many of these issues are inherently political, and some involve public goods; in those cases, live and let live is not always a viable solution. If a particular platform enables negative externalities, or threatens to push society into a suboptimal equilibrium, then you cannot "opt out" simply by using your platform instead. There, some kind of network-effect-driven or even in extreme cases 51% attack–driven censure may be necessary.* In some cases, the differences are related to private goods, and are primarily simply a matter of empirical beliefs. If I believe that SchellingDollar is the best scheme for price stability, and others prefer Seigniorage Shares or NuBits, then after a few years or decades one model will prove to work better, replace its competition, and that will be that.

In other cases, however, the differences will be resolved in a different way: it will turn out that the properties of some systems are better suited for some applications, and other systems better suited for other applications, and everything will naturally specialize into those use cases where it works best. As a number of commentators have pointed out, for decentralized consensus applications in the mainstream financial world, banks will likely not be willing to accept a network managed by anonymous nodes; in this case, something like Ripple will be more useful. But for Silk Road 4.0, the exact opposite approach is the only way to go—and for everything in between it's a cost-benefit analysis all the way. If users want networks specialized to performing specific functions highly efficiently, then networks will exist for that, and if users want a general-purpose network with a high network

---

* A 51% attack is the dreaded event wherein a miner gains majority control over a blockchain network and has the ability to falsify transactions.

effect between on-chain applications then that will exist as well. As David Johnston points out, blockchains are like programming languages: they each have their own particular properties, and few developers religiously adhere to one language exclusively—rather, we use each one in the specific cases for which it is best suited.

ROOM FOR COOPERATION

However, as was mentioned earlier, this does not mean that we should simply go our own way and try to ignore—or worse, actively sabotage—each other. Even if all of our projects are necessarily specializing toward different goals, there is nevertheless a substantial opportunity for much less duplication of effort, and more cooperation. This is true on multiple levels. First, let us look at a model of the cryptocurrency ecosystem—or, perhaps, a vision of what it might look like in one to five years' time:

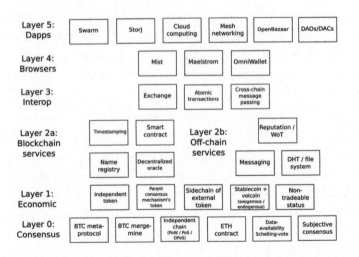

Ethereum has its own presence on pretty much every level:

- Consensus: Ethereum blockchain, data-availability Schelling-vote (maybe for Ethereum 2.0)

- Economics: ether, an independent token, as well as research into stablecoin proposals

- Blockchain services: name registry

- Off-chain services: Whisper (messaging), web of trust (in progress)

- Interop: BTC-to-ether bridge (in progress)

- Browsers: Mist

Now, consider a few other projects that are trying to build holistic ecosystems of some kind.

BitShares has at the least:

- Consensus: DPoS

- Economics: BTSX and BitAssets

- Blockchain services: BTS decentralized exchange

- Browsers: BitShares client (though not quite a browser in the same concept)

MaidSafe has:

- Consensus: SAFE Network

- Economics: Safecoin

- Off-chain services: Distributed hash table, MaidSafe Drive

BitTorrent has announced their plans for Maelstrom, a project intended to serve a rather similar function to Mist, albeit showcasing their own (not blockchain-based) technology. Cryptocurrency projects generally all build a blockchain, a currency, and a client of their own, although forking a single client is common for the less innovative cases. Name-registration and identity-management systems are now a dime a dozen. And, of course, just about every project realizes that it has a need for some kind of reputation and web of trust.

Now, let us paint a picture of an alternative world. Instead of having a collection of cleanly disjoint, vertically-integrated ecosystems, with each one building its own components for everything, imagine a world where Mist could be used to access Ethereum, BitShares, MaidSafe, or any other major decentralized-infrastructure network, with new decentralized networks being installable much like the plugins for Flash and Java inside of Chrome and Firefox. Imagine that the reputation data in the web of trust for Ethereum could be reused in other projects as well. Imagine Storj running inside of Maelstrom as a dapp,* using MaidSafe for a file-storage backend, and using the Ethereum blockchain to maintain the contracts that incentivize continued storage and downloading. Imagine identities being automatically transferrable across any crypto-networks, as long as they use the same underlying cryptographic algorithms (e.g., ECDSA + SHA3).

The key insight here is this: although some of the layers in the ecosystem are inextricably linked—for example, a single dapp will often correspond to a single specific service on the Ethereum blockchain—in many cases the layers can easily be designed to be much more modular, allowing each product on each layer to compete separately on its own merits. Browsers are perhaps the

---

* The term "dapp" simply means "decentralized application," referring to software built to run on a blockchain rather than on someone's server. The projects referred to here are early efforts to build such software.

most separable component; most reasonably holistic lower-level blockchain service sets have similar needs in terms of what applications can run on them, and so it makes sense for each browser to support each platform. Off-chain services are also a target for abstraction; any decentralized application, regardless of what blockchain technology it uses, should be free to use Whisper, Swarm, IPFS, or any other service that developers come up with. On-chain services, like data provision, can theoretically be built so as to interact with multiple chains.

Additionally, there are plenty of opportunities to collaborate on fundamental research and development. Discussion on proof of work, proof of stake, stable currency systems, and scalability, as well as other hard problems of cryptoeconomics, can easily be substantially more open, so that the various projects can benefit from and be more aware of each other's developments. Basic algorithms and best practices related to networking layers, cryptographic algorithm implementations, and other low-level components can, and should, be shared. Interoperability technologies should be developed to facilitate easy exchange and interaction between services and decentralized entities on one platform and another. The Cryptocurrency Research Group is one initiative that we plan to initially support, with the hope that it will grow to flourish independently of ourselves, with the goal of promoting this kind of cooperation. Other formal and informal institutions can doubtlessly help support the process.

Hopefully, in the future we will see many more projects existing in a much more modular fashion, living on only one or two layers of the cryptocurrency ecosystem, and providing a common interface allowing any mechanism on any other layer to work with them. If the cryptocurrency space goes far enough, then even Firefox and Chrome may end up adapting themselves to process decentralized-application protocols as well. A journey toward such an ecosystem is not something that needs to be rushed imme-diately; at this point, we have quite little idea of what kinds of blockchain-driven services people will be using in the first place, making it hard to determine exactly what kind of interoperability would actually be useful. However, things slowly but surely are taking their first few steps in that direction; Eris's Decerver, their own "browser" into the decentralized world, supports access to Bitcoin, Ethereum, and their own Thelonious blockchains, as well as an IPFS content hosting network.

There is room for many projects that are currently in the crypto 2.0 space to succeed, and so having a winner-take-all mentality at this point is completely unnecessary and harmful. All that we need to do right now to set off on the journey on a better road is to live with the assumption that we are all building our own platforms, tuned to our own particular set of preferences and parameters, but at the end of the day a plurality of networks will succeed and we will need to live with that reality, so we might as well start preparing for it now.

Happy new year, and looking forward to an exciting 2015 007 Anno Satoshii.

SUPERRATIONALITY AND DAOS

Ethereum blog
January 23, 2015

One of the common questions that many in the crypto 2.0 space have about the concept of decentralized autonomous organizations is a simple one: What are DAOs good for? What fundamental advantage would an organization have from its management and operations being tied down to hard code on a public blockchain that could not be had by going the more traditional route? What advantages do blockchain contracts offer over plain old shareholder agreements? Particularly, even if public-good rationales in favor of transparent governance, and guaranteed-not-to-be-evil governance can be raised, what is the incentive for an individual organization to voluntarily weaken itself by opening up its innermost source code, where its competitors can see every single action that it takes or even plans to take while operating behind closed doors?

There are many paths that one could take to answering this question. For the specific case of nonprofit organizations that are already explicitly dedicating themselves to charitable causes, one can rightfully say that they lack individual incentive; they are already dedicating themselves to improving the world for little or no monetary gain to themselves. For private companies, one can make the information-theoretic argument that a governance algorithm will work better if, all else being equal, everyone can participate and introduce their own information and intelligence into the calculation—a rather reasonable hypothesis given the established result from machine learning that much larger performance gains can be made by increasing the data size than by tweaking the algorithm. In this article, however, we will take a different and more specific route.

WHAT IS SUPERRATIONALITY?

In game theory and economics, it is a very widely understood result that there exist many classes of situations in which a set of individuals have the opportunity to act in one of two ways, either "cooperating" with or "defecting" against each other, such that everyone would be better off if everyone cooperated, but regardless of what others do each individual would be better off by themselves defecting. As a result, the story goes, everyone ends up defecting, and so people's individual rationality leads to the worst possible collective result. The most common example of this is the celebrated prisoner's dilemma game.

Since many readers have likely already seen the prisoner's dilemma, I will spice things up by giving Eliezer Yudkowsky's rather deranged version of the game:

> Let's suppose that four billion human beings—not the whole human species, but a significant part of it—are currently progressing through a fatal disease that can only be cured by substance S.
>
> However, substance S can only be produced by working with [a strange AI from another dimension whose only

goal is to maximize the quantity of paperclips]—substance S can also be used to produce paperclips. The paperclip maximizer only cares about the number of paperclips in its own universe, not in ours, so we can't offer to produce or threaten to destroy paperclips here. We have never interacted with the paperclip maximizer before, and will never interact with it again. Both humanity and the paperclip maximizer will get a single chance to seize some additional part of substance S for themselves, just before the dimensional nexus collapses; but the seizure process destroys some of substance S.

The payoff matrix is as follows:

	Humans cooperate	Humans defect
AI cooperates	2 billion lives saved, 2 paperclips gained	3 billions lived saved, 0 paperclips gained
AI defects	0 lives saved, 3 paperclips gained	1 billion lives saved, 1 paperclip gained

From our point of view, it obviously makes sense from a practical, and in this case moral, standpoint that we should defect; there is no way that a paperclip in another universe can be worth a billion lives. From the AI's point of view, defecting always leads to one extra paperclip, and its code assigns a value to human life of exactly zero; hence, it will defect. However, the outcome that this leads to is clearly worse for both parties than if the humans and AI both cooperated—but then, if the AI was going to cooperate, we could save even more lives by defecting ourselves, and likewise for the AI if we were to cooperate.

In the real world, many two-party prisoner's dilemmas on the small scale are resolved through the mechanism of trade and the ability of a legal system to enforce contracts and laws; in this case, if there existed a god who has absolute power over both universes but cared only about compliance with one's prior agreements, the humans and the AI could sign a contract to cooperate and ask the god to simultaneously prevent both from defecting. When there is no ability to precontract, laws penalize unilateral defection. However, there are still many situations, particularly when many parties are involved, where opportunities for defection exist:

□ Alice is selling lemons in a market, but she knows that her current batch is low quality and once customers try to use them they will immediately have to throw them out. Should she sell them anyway? (Note that this is the sort of marketplace where there are so many sellers you can't really keep track of reputation). Expected gain to Alice: $5 revenue per lemon − $1 shipping and store costs = $4. Expected cost to society: $5 revenue − $1 costs − $5 wasted money from customer = -$1. Alice sells the lemons.

□ Should Bob donate $1,000 to Bitcoin development? Expected gain to society: $10 × 100,000 people − $1,000 = $999,000. Expected gain to Bob: $10 − $1000 = -$990, so Bob does not donate.

□ Charlie found someone else's wallet, containing $500. Should he return it? Expected gain to society: $500 (to recipient) − $500 (Charlie's loss) + $50 (intangible gain to society from everyone being able to worry a little less about the safety of their wallets). Expected gain to Charlie: -$500, so he keeps the wallet.

□ Should David cut costs in his factory by dumping toxic waste into a river? Expected gain to society: $1,000 savings

– $10 average increased medical costs × 100,000 people = -$999,000, expected gain to David: $1,000 − $10 = $990, so David pollutes.

□ Eve developed a cure for a type of cancer which costs $500 per unit to produce. She can sell it for $1,000, allowing 50,000 cancer patients to afford it, or for $10,000, allowing 25,000 cancer patients to afford it. Should she sell at the higher price? Expected gain to society: -25,000 lives (including Eve's profit, which cancels out the wealthier buyers' losses). Expected gain to Eve: $237.5 million profit instead of $25 million = $212.5 million, so Eve charges the higher price.

Of course, in many of these cases, people sometimes act morally and cooperate, even though it reduces their personal situation. But why do they do this? We were produced by evolution, which is generally a rather selfish optimizer. There are many explanations. One, and the one we will focus on, involves the concept of superrationality.

SUPERRATIONALITY

Consider the following explanation of virtue, courtesy of David Friedman:

I start with two observations about human beings. The first is that there is a substantial connection between what goes on inside and outside of their heads. Facial expressions, body positions, and a variety of other signs give us at least some idea of our friends' thoughts and emotions. The second is that we have limited intellectual ability—we cannot, in the time available to make a decision, consider all options. We are, in the jargon of computers, machines of limited computing power operating in real time. Suppose I wish people to believe that I have certain characteristics—that I am honest, kind, helpful to my friends. If I really do have those characteristics, projecting them is easy—I merely do and say what seems natural, without paying much attention to how I appear to outside observers. They will observe my words, my actions, my facial expressions, and draw reasonably accurate conclusions. Suppose, however, that I do not have those characteristics. I am not (for example) honest. I usually act honestly because acting honestly is usually in my interest, but I am always willing to make an exception if I can gain by doing so. I must now, in many actual decisions, do a double calculation. First, I must decide how to act—whether, for example, this is a good opportunity to steal and not be caught. Second, I must decide how I would be thinking and acting, what expressions would be going across my face, whether I would be feeling happy or sad, if I really were the person I am pretending to be. If you require a computer to do twice as many calculations, it slows down. So does a human. Most of us are not very good liars. If this argument is correct, it implies that I may be better off in narrowly material terms—have, for instance, a higher income—if I am really honest (and kind and . . .) than if I am only pretending to be, simply because real virtues are more convincing than pretend ones. It follows that, if I were a narrowly selfish individual, I might, for purely selfish reasons, want to make myself a better person—more virtuous in those ways that others value. The final stage in the argument is to observe that we can be made better—by ourselves, by our parents, perhaps even by our genes. People can and do try to train themselves into good habits—including the habits of automatically telling the truth, not stealing, and being kind to their friends. With enough training, such habits become tastes—doing "bad" things makes one uncomfortable, even if nobody is watching, so one does not do them. After a while, one does not even have to decide not to do them. You might describe the process as synthesizing a conscience.

Essentially, it is cognitively hard to convincingly fake being virtuous while being greedy whenever you can get away with it, and so it makes more sense for you to actually be virtuous. Much ancient philosophy follows similar reasoning, seeing virtue as a cultivated habit; David Friedman simply did us the customary service of an economist and converted the intuition into more easily analyzable formalisms. Now, let us compress this formalism even further. In short, the key point here is that humans are leaky agents—with every second of our action, we essentially indirectly expose parts of our source code. If we are actually planning to be nice, we act one way, and if we are only pretending to be nice while actually intending to strike as soon as our friends are vulnerable, we act differently, and others can often notice.

This might seem like a disadvantage; however, it allows a kind of cooperation that was not possible with the simple game-theoretic agents described above. Suppose that two agents, A and B, each have the ability to "read" whether or not the other is "virtuous" to some degree of accuracy, and are playing a symmetric prisoner's dilemma. In this case, the agents can adopt the following strategy, which we assume to be a virtuous strategy:

1. Try to determine if the other party is virtuous.

2. If the other party is virtuous, cooperate.

3. If the other party is not virtuous, defect.

If two virtuous agents come into contact with each other, both will cooperate, and get a larger reward. If a virtuous agent comes into contact with a non-virtuous agent, the virtuous agent will defect. Hence, in all cases, the virtuous agent does at least as well as the non-virtuous agent, and often better. This is the essence of superrationality.

As contrived as this strategy seems, human cultures have some deeply ingrained mechanisms for implementing it, particularly relating to mistrusting agents who try hard to make themselves less readable—see the common adage that you should never trust someone who doesn't drink. Of course, there is a class of individuals who can convincingly pretend to be friendly while actually planning to defect at every moment—these are called sociopaths, and they are perhaps the primary defect of this system when implemented by humans.

CENTRALIZED MANUAL ORGANIZATIONS . . .

This kind of superrational cooperation has been arguably an important bedrock of human cooperation for the last ten thousand years, allowing people to be honest to each other even in those cases where simple market incentives might instead drive defection. However, perhaps one of the main unfortunate byproducts of the modern birth of large centralized organizations is that they allow people to effectively cheat others' ability to read their minds, making this kind of cooperation more difficult.

Most people in modern civilization have benefited quite handsomely, and have also indirectly financed, at least some instance of someone in some third-world country dumping toxic waste into a river to build products more cheaply for them; however, we do not even realize that we are indirectly participating in such defection; corporations do the dirty work for us. The market is so powerful that it can arbitrage even our own morality, placing the most dirty and unsavory tasks in the hands of those individuals who are willing to absorb their conscience at lowest cost and effectively hiding it from everyone else. The corporations themselves are perfectly able to have a smiley face produced as their public image by their marketing departments, leaving it to a completely different department to sweet-talk potential customers. This second department may not even know that the department producing the product is any less virtuous and sweet than they are.

The internet has often been hailed as a solution to many of these organizational and political problems, and indeed it does a great job of reducing information asymmetries and offering transparency. However, as far as the decreasing viability of superrational cooperation goes, it can also sometimes make things even worse. Online, we are much less "leaky" even as individuals, and so once again it is easier to appear virtuous while actually intending to cheat. This is part of the reason why scams online and in the cryptocurrency space are more common than they are offline, and is perhaps one of the primary arguments against moving all economic interaction to the internet à la crypto anarchism (the other argument being that crypto anarchism removes the ability to inflict unboundedly large punishments, weakening the strength of a large class of economic mechanisms).

A much greater degree of transparency, arguably, offers a solution. Individuals are moderately leaky, current centralized organizations are less leaky, but organizations where information is constantly, randomly being released to the world left, right, and center are even more leaky than individuals are. Imagine a world where if you start even thinking about how you will cheat your friend, business partner, or spouse, there is a 1% chance that the left part of your hippocampus will rebel and send a full recording of your thoughts to your intended victim in exchange for a $7,500 reward. That is what it "feels" like to be the management board of a leaky organization.

This is essentially a restatement of the founding ideology behind WikiLeaks, and more recently an incentivized WikiLeaks alternative, slur.io, came out to push the envelope further. However, WikiLeaks exists, and yet shadowy centralized organizations also continue to still exist and are in many cases still quite shadowy. Perhaps incentivization, coupled with prediction-like mechanisms for people to profit from outing their employers' misdeeds, is what will open the floodgates for greater transparency, but at the same time we can also take a different route: offer a way for organizations to make themselves voluntarily, and radically, leaky and superrational to an extent never seen before.

. . . AND DAOS

Decentralized autonomous organizations, as a concept, are unique in that their governance algorithms are not just leaky, but actually completely public. That is, while with even transparent centralized organizations outsiders can get a rough idea of what the organization's temperament is, with a DAO outsiders can actually see the organization's entire source code. Now, they do not see the "source code" of the humans that are behind the DAO, but there are ways to write a DAO's source code so that it is heavily biased toward a particular objective regardless of who its participants are.

A futarchy maximizing the average human lifespan will act very differently from a futarchy maximizing the production of paperclips, even if the exact same people are running it. Hence, not only is it the case that the organization will make it obvious to everyone if they start to cheat, it's not even possible for the organization's "mind" to cheat.

Now, what would superrational cooperation using DAOs look like? First, we would need to see some DAOs actually appear. There are a few use-cases where it seems not too farfetched to expect them to succeed: gambling, stablecoins, decentralized file storage, one-ID-per-person data provision, SchellingCoin, etc. However, we can call these DAOs "type I DAOs": they have some internal state, but little autonomous governance. They cannot ever do anything but perhaps adjust a few of their own parameters to maximize some utility metric via PID controllers, simulated annealing, or other simple optimization algorithms. Hence, they are in a weak sense superrational, but they are also rather limited and stupid, and so they will often rely on being upgraded by an external process which is not superrational at all.

In order to go further, we need type II DAOs: DAOs with a governance algorithm capable of making theoretically arbitrary decisions. Futarchy, various forms of democracy, and various forms of subjective extra-protocol governance (i.e., in case of substantial disagreement, DAO clones itself into multiple parts with one part for each proposed policy, and everyone chooses which version to interact with) are the only ones we are currently aware of, though other fundamental approaches and clever combinations of these will likely continue to appear. Once DAOs can make arbitrary decisions, then they will be able to not only engage in superrational commerce with their human customers but also potentially with each other.

What kinds of market failures can superrational cooperation

solve that plain old regular cooperation cannot? Public-goods problems may unfortunately be outside the scope; none of the mechanisms described here solve the massively-multiparty incentivization problem. In this model, the reason why organizations make themselves decentralized/leaky is so that others will trust them more, and so organizations that fail to do this will be excluded from the economic benefits of this "circle of trust." With public goods, the whole problem is that there is no way to exclude anyone from benefiting, so the strategy fails. However, anything related to information asymmetries falls squarely within the scope, and this scope is large indeed; as society becomes more and more complex, cheating will in many ways become progressively easier and easier to do and harder to police or even understand; the modern financial system is just one example. Perhaps the true promise of DAOs, if there is any promise at all, is precisely to help with this.

THE VALUE OF BLOCKCHAIN TECHNOLOGY

Ethereum blog
April 13, 2015

One of the questions that has perhaps been central to my own research in blockchain technology is: Ultimately, what is it even useful for? Why do we need blockchains for anything, what kinds of services should be run on blockchain-like architectures, and why specifically should services be run on blockchains instead of just living on plain old servers? Exactly how much value do blockchains provide: are they absolutely essential, or are they just nice to have? And, perhaps most importantly of all, what is the "killer app" going to be?

Over the last few months, I have spent a lot of time thinking about this issue, discussing it with cryptocurrency developers, venture capital firms, and particularly people from outside the blockchain space, whether civil liberties activists or people in the finance and payments industry or anywhere else. In the process, I have come to a number of important, and meaningful, conclusions.

First, there will be no "killer app" for blockchain technology. The reason for this is simple: the doctrine of low-hanging fruit.

If there existed some particular application for which blockchain technology is massively superior to anything else for a significant portion of the infrastructure of modern society, then people would be loudly talking about it already. This may seem like the old economics joke about an economist finding a twenty-dollar bill on the ground and concluding it must be fake, because otherwise it would already have been taken, but in this case the situation is subtly different: unlike the dollar bill, where search costs are low and so picking up the bill makes sense even if there is only a 0.01% chance it is real, here search costs are very high, and plenty of people with billions of dollars of incentive have already been searching. And so far, there has been no single application that anyone has come up with that has seriously stood out to dominate everything else on the horizon.

In fact, one can quite reasonably argue that the closest things that we will ever have to "killer apps" are precisely those apps that have already been done and recited and sensationalized ad nauseam: censorship resistance for WikiLeaks and Silk Road. Silk Road, the online anonymous drug marketplace that was shut down by law enforcement in late 2013, processed over $1 billion in sales during its two and a half years of operations, and while the payment-system-orchestrated blockade against WikiLeaks was in progress, Bitcoin and Litecoin donations were responsible for the bulk of its revenue.* In both cases the need was clear and the potential economic surplus was very high—before Bitcoin, you would have no choice but to buy the drugs in person and donate to WikiLeaks by cash-in-the-mail, and so Bitcoin provided a massive convenience gain, and thus the opportunity was snatched up almost instantly. Now, however, that is much less the case, and

---

* After the website WikiLeaks released leaked documents relating to the wars in Iraq and Afghanistan in 2010, the US government orchestrated a withdrawal of financial services from the organization. The following year, WikiLeaks enabled Bitcoin donations.

marginal opportunities in blockchain technology are not nearly such easy grabs.

TOTAL AND AVERAGE UTILITY

Does this mean, however, that blockchains have hit their peak utility? Most certainly not. They have hit peak necessity, in the sense of *peak utility per user*, but that is not the same thing as peak utility. Although Silk Road was indispensable for many of the people that used it, even among the drug-using community it's not indispensable in general; as much as it befuddles this particular author how ordinary individuals are supposed to get such connections, most people have somehow found "a guy" that they know that they can purchase their weed from. Interest in smoking weed at all seems to strongly correlate with having easy access to it. Hence, in the grand scheme of things, Silk Road has only had a chance to become relevant to a very niche group of people. WikiLeaks is similar; the set of people who care about corporate and governmental transparency strongly enough to donate money to a controversial organization in support of it is not very large compared to the entire population of the world. So what's left? In short, the long tail.

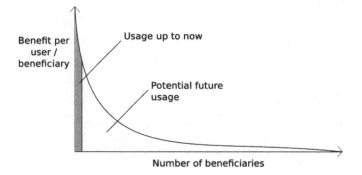

So what is the long tail? This is where it gets hard to explain. I could provide a list of applications that are included in this "long tail"; however, blockchains are not indispensable and do not even offer extremely strong fundamental advantages for each one. For each individual case, an advocate of either the "blockchain applications are overrated, it's the Bitcoin currency that matters" or the "blockchain tech as a whole is useless" position can quite reasonably come up with a way to implement the scheme just as easily on a centralized server, replace blockchain governance with a legal contract, and apply whatever other replacements to turn the product into something much more similar to a traditional system. And on that point, they would be completely correct: for that particular use case, blockchains are not indispensable. And that's the whole point: those applications are not at the top of the distribution, up there with WikiLeaks and Silk Road; if they were, they would have been implemented already. In the long tail, blockchains are not necessary; *they are convenient.* They are simply marginally better than the next available tool for the job. And yet, because these applications are much more mainstream, and can benefit hundreds of millions of users, the total gain to society (which can be seen from the area on the above chart) is much larger.

Perhaps the best analogy to this line of reasoning is to ask the following rhetorical question: What is the killer app of "open source"? Open source has clearly been a very good thing for society, and it is being used for millions of software packages around the world, but nevertheless it is still hard to answer the question. And the reason is the same: there is no killer app, and the list of applications has a very, very long tail—basically, just about every kind of software imaginable, with particular emphasis on lower-level libraries that end up reused by millions of projects many times over and critical cryptographic security libraries.

BLOCKCHAINS, REDEFINED . . . AGAIN

Now, what are the specific benefits of blockchains that make the long tail worthwhile? To start off, let me provide the current description that I use of what a blockchain is:

> A blockchain is a magic computer that anyone can upload programs to and leave the programs to self-execute, where the current and all previous states of every program are always publicly visible, and which carries a very strong cryptoeconomically secured guarantee that programs running on the chain will continue to execute in exactly the way that the blockchain protocol specifies.

Notice that this definition does NOT:

☐ use financially charged terms like "ledger," "money," or "transactions," or indeed any terms geared toward a particular use case;

☐ mention any particular consensus algorithm, or indeed mention anything about the technical properties of how a blockchain works (except for the fact that it's "cryptoeconomic," a technical term roughly meaning "it's decentralized, it uses public key cryptography for authentication, and it uses economic incentives to ensure that it keeps going and doesn't go back in time or incur any other glitch");

☐ make a restriction to any particular type of state transition function.

The one thing that the definition does well is explain what a blockchain does, and it explains it in such a way that any software developer will be able to fairly clearly have at least an intuitive grasp of its value proposition. Now, in practice, sometimes the programming language that the programs run in is very restrictive; Bitcoin's language can be seen as requiring a sequence of `DESTROY COIN: <txid> <index> <scriptsig>` statements followed by a sequence of `CREATE COIN: <scriptpubkey> <value>` statements, where `scriptpubkey` is a restricted mathematical formula, `scriptsig` must be a satisfying variable assignment to the formula (e.g., $\{x = 5, \ y = 7\}$ satisfies $2 \times x - y = 3$), and an attempt to destroy a nonexistent coin or destroy a coin without supplying a valid `scriptsig` for that coin's `scriptpubkey`, or an attempt to create more coin value than you destroyed, returns an error. Other programming languages, on the other hand, can be much more expressive. It's up to the software developer to analyze what programming language is right for their task, much like it is a software developer's task today to decide between Python, C++, Node.js, and Malbolge.

The one thing that the definition emphasizes extremely well is that blockchains are not about bringing to the world any one particular ruleset, whether it's a currency with a fixed-supply monetary policy, a name registry with a two-hundred-day reregistration time, a particular decentralized exchange design, or whatever else; rather, they're about creating the freedom to create a new mechanism with a new ruleset extremely quickly and pushing it out. They're Lego Mindstorms for building economic and social institutions.

This is the core of the more moderate version of the "it's the blockchain that's exciting, not the currency" position that is so prevalent in mainstream industry: it is indeed true that currency is necessary to make cryptoeconomic blockchains work (although NOT blockchain-like data structures following the Stellar subjective consensus model), but the currency is there simply as economic plumbing to incentivize consensus participation, hold deposits, and pay transaction fees, not as the center-stage point of speculative mania, consumer interest and excitement.

Now, why are blockchains useful? To summarize:

☐ You can store data on them and that data is guaranteed to have a very high degree of availability.

☐ You can run applications on them and be guaranteed an extremely high uptime.

☐ You can run applications on them, and be guaranteed an extremely high uptime going very far into the future.

☐ You can run applications on them, and convince your users that the application's logic is honest and is doing what you are advertising that it does.

☐ You can run applications on them, and convince your users that your application will remain working even if you lose interest in maintaining it, you are bribed or threatened to manipulate the application state in some way, or you acquire a profit motive to manipulate the application state in some way.

☐ You can run applications on them, and give yourself the backdoor key if it is absolutely necessary, BUT put "constitutional" limitations on your use of the key—for example, requiring a software update to pass through a public one-month waiting period before it can be introduced, or at the very least immediately notifying users of application updates.

☐ You can run applications on them, and give a backdoor

key to a particular governance algorithm (e.g., voting, futarchy, some complicated multicameral parliament architecture), and convince your users that the particular governance algorithm in question is actually in control of the application.

☐ You can run applications on them, and those applications can talk to each other with 100% reliability—even if the underlying platform has only 99.999% reliability.

☐ Multiple users or companies can run applications on them, and those applications can interact with each other at extremely high speed without requiring any network messages, while at the same time ensuring that each company has total control over its own application.

☐ You can build applications that very easily and efficiently take advantage of the data produced by other applications. (Combining payments and reputation systems is perhaps the largest gain here.)

All of those things are valuable indirectly to billions of people around the world, potentially particularly in regions of the world where highly developed economic, financial, and social infrastructure currently does not work at all (though technology will often need to be combined with political reforms to solve many of the problems), and blockchains are good at providing these properties. They are obviously valuable in finance, as finance is perhaps the most simultaneously computationally and trust-intensive industry in the world, but they are also valuable in many other spots in internet infrastructure. There do exist other architectures that can also provide these properties, but they are *slightly to moderately less good* than blockchains are. Gavin Wood has started describing this ideal computing platform as "the world computer"—a computer the state of which is shared among everyone and which a very large group of people, which anyone is free to join, are involved in maintaining.

BASE-LAYER INFRASTRUCTURE

Like open source, by far the largest opportunity for gains out of blockchain technology are out of what can be called "base-layer infrastructure" services. Base-layer infrastructure services, as a general category, are characterized by the following properties:

□ Dependency—there exist many other services that intimately depend on the base-layer service for functionality;

□ High network effects—there are substantial benefits from very large groups of people (or even everyone) using the same service;

□ High switching costs—it is difficult for an individual to switch from one service to the other.

Note that one concern that is not in there is any notion of raw "necessity" or "importance"; there can be fairly unimportant base layers (e.g., RSS feeds) and important non-base layers (e.g., food). Base-layer services have existed ever since even before the dawn of civilization; in the so-called "caveman days" the single most important base-layer service of all was language. In somewhat more recent times, the primary examples became roads, the legal system, and the postal and transportation systems; in the twentieth century we added telephone networks and financial systems, and at the end of the millennium emerged the internet. Now, however, the new base-layer services of the internet are almost entirely informational: internet payment systems, identity, domain-name systems, certificate authorities, reputation systems, cloud computing, various kinds of data feeds, and perhaps in the near future prediction markets.

In ten years' time, the highly networked and interdependent nature of these services may make it such that it is harder for individuals to switch from one system to another than it is for them to even switch which government they are living under—and that means that making sure that these services are built correctly and that their governance process does not put a few private entities in positions of extreme power is of utmost importance. Right now, many of these systems are built in a highly centralized fashion, and this is in part simply due to the fact that the original design of the World Wide Web failed to realize the importance of these services and include defaults—and so, even today, most websites ask you to "sign in with Google" or "sign in with Facebook," and certificate authorities run into problems like this:*

☐   A solo Iranian hacker on Saturday claimed responsibility for stealing multiple SSL certificates belonging to some of the web's biggest sites, including Google, Microsoft, Skype, and Yahoo.

☐   Early reaction from security experts was mixed, with some believing the hacker's claim, while others were dubious.

☐   Last week, conjecture had focused on a state-sponsored attack, perhaps funded or conducted by the Iranian government, that hacked a certificate reseller affiliated with US-based Comodo.

☐   On March 23, Comodo acknowledged the attack, saying that eight days earlier, hackers had obtained nine bogus

---

* The following is quoted from a 2011 *Computerworld* article by Gregg Keizer.

certificates for the log-on sites of Microsoft's Hotmail, Google's Gmail, the internet phone and chat service Skype, and Yahoo Mail. A certificate for Mozilla's Firefox add-on site was also acquired.

Why shouldn't certificate authorities be decentralized at least to the point of an M-of-N system* again? (Note that the case for much more widespread use of M-of-N is logically separable from the case for blockchains, but blockchains happen to be a good platform to run M-of-N on.)

IDENTITY

Let us take a particular use case, "identity on the blockchain," and run with it. In general, what do you need in order to have an identity? The simplest answer is one we already know: you need to have a public and private key. You publish the public key, which becomes your ID, and you digitally sign every message you send with your private key, allowing anyone to verify that those messages were produced by you (where, from their point of view, "you" means "the entity that holds that particular public key"). However, there are a few challenges:

1. What happens if your key gets stolen, and you need to switch to a new one?

2. What happens if you lose your key?

3. What if you want to refer to other users by their names, and not just a random twenty-byte string of cryptographic data?

---

* An M-of-N system is one in which, for instance, there are some number N keys to a lock and, of those, M keys are needed to unlock it.

4. What if you want to use a more advanced approach for security such as multisig, and not just a single key?

Let us try solving these challenges one by one. We can start off with the fourth. A simple solution is this: instead of requiring one particular cryptographic signature type, your public key becomes a program, and a valid signature becomes a string that, when fed into the program together with the message, returns 1. Theoretically, any single-key, multi-key, or whatever other kind of ruleset can be encoded into such a paradigm.

However, this has a problem: the public keys will get too long. We can solve this by putting the actual "public key" into some data store (e.g., a distributed hash table if we want decentralization) and using the hash of the "public key" as the user's ID. This does not yet require blockchains—although, in the latest designs, scalable blockchains are really not that different from DHTs and so it is entirely possible that, in ten years' time, every kind of decentralized system used for anything will accidentally or intentionally converge into some kind of scalable blockchain.

Now, consider the first problem. We can think of this as the certificate revocation problem: If you want to "revoke" a particular key, how do you ensure that it gets around to everyone who needs to see it? This by itself can once again be solved by a distributed hash table. However, this leads to the next problem: If you want to revoke a key, what do you replace it with? If your key is stolen, you and the attacker both have it, and so neither of you can be convincingly more authoritative. One solution is to have three keys, and then if one gets revoked, require a signature from two or all of them to approve the next key. But this leads to a "nothing at stake" problem: if the attacker *eventually* manages to steal all three of your keys from some point in history, then they can simulate a history of assigning a new key, assigning further new keys from there, and your own history is no longer more authoritative. *This* is a timestamping problem, and so here blockchains can actually help.

For the second problem, holding multiple keys and reassigning also works reasonably well—and here, blockchains are not needed. In fact, you do not need to reassign; with clever use of secret sharing you can actually recover from key losses simply by keeping your key in "shards," such that if you lose any single shard you can always use secret-sharing math to simply recover it from the others. For the third problem, blockchain-based name registries are the simplest solution.

However, in practice most people are not well-equipped to securely store multiple keys, and there are always going to be mishaps, and often centralized services play an important role: helping people get their accounts back in the event of a mistake. In this case, the blockchain-based solution is simple: social M-of-N backup.

You pick eight entities; they may be your friends, your employer, some corporation, a nonprofit, or even in the future a government, and if anything goes wrong a combination of five of them can recover your key. This concept of social multisignature backup is perhaps one of the most powerful mechanisms to use in any kind of decentralized system design, and provides a very high amount of security very cheaply and without relying on centralized trust. Note that blockchain-based identity, particularly with Ethereum's contract model, makes all of this very easy to program: in the name registry, register your name and point it at a contract, and have that contract maintain the current main key and backup keys associated with the identity as well as the logic for updating them over time. An identity system, safe and easy-to-use enough for grandma, done without any individual entity (except for you!) in control.

Identity is not the only problem that blockchains can alleviate. Another component, intimately tied up with identity, is reputation. Currently, what passes for "reputation systems" in the modern world are invariably either insecure, due to their inability to ensure that an entity rating another entity actually interacted with them, or centralized, tying reputation data to a particular platform and having the reputation data exist under that platform's control. When you switch from Uber to Lyft, your Uber rating does not carry over.

A decentralized reputation system would ideally consist of two separate layers: data and evaluation. Data would consist of individuals making independent ratings about others, ratings tied to transactions (e.g., with blockchain-based payments one can create an open system such that you can only give merchants a rating if you actually pay them), and a collection of other sources, and anyone can run their own algorithm to evaluate their data; "light-client friendly" algorithms that can evaluate a proof of reputation from a particular dataset quickly may become an important research area (many naïve reputation algorithms involve matrix math, which has nearly cubic computational complexity in the underlying data and so is hard to decentralize). "Zero-knowledge" reputation systems that allow a user to provide some kind of cryptographic certificate proving that they have at least x reputation points according to a particular metric without revealing anything else are also promising.

The case of reputation is interesting because it combines together multiple benefits of the blockchain as a platform:

☐ Its use as a data store for identity

☐ Its use as a data store for reputational records

☐ Inter-application interoperability (ratings tied to proof of payment, ability for any algorithm to work over the same underlying set of data, etc.)

□   A guarantee that the underlying data will be portable going into the future (companies may voluntarily provide a reputation certificate in an exportable format, but they have no way to pre-commit to continuing to have that functionality going into the future)

□   The use of a decentralized platform more generally to guarantee that the reputation wasn't manipulated at the point of calculation

Now, for all of these benefits, there are substitutes: we can trust Visa and Mastercard to provide cryptographically signed receipts that a particular transaction took place, we can store reputational records on archive.org, we can have servers talk to each other, we can have private companies specify in their terms of service that they agree to be nice, and so forth. All of these options are reasonably effective, but they are *not nearly as nice* as simply putting everything out into the open, running it on the "world computer," and letting cryptographic verification and proofs do the work. And a similar argument can be made for every other use case.

CUTTING COSTS

If the largest value from blockchain technology comes at the long tail, as this thesis suggests, then that leads to an important conclusion: *the per-transaction gain from using a blockchain is very small*. Hence, the problem of cutting costs of consensus and increasing blockchain scalability becomes paramount. With centralized solutions, users and businesses are used to paying essentially zero dollars per "transaction"; although individuals looking to donate to WikiLeaks may be willing to pay even a fee of $5 to get their transaction through, someone trying to upload a reputation record may well only be willing to pay a fee of $0.0005.

Hence, the problem of making consensus cheaper, both in the absolute sense (i.e., proof of stake) and in the per-transaction sense (i.e., through scalable blockchain algorithms, where at most a few hundred nodes process each transaction), is absolutely paramount. Additionally, blockchain developers should keep in mind that the last forty years of software development has been a history of moving to progressively less and less efficient programming languages and paradigms solely because they allow developers to be less experienced and lazier. It is necessary to design blockchain algorithms that incorporate the principle that developers are really not going to be all that smart and judicious about what they put on the blockchain and what they keep off—though a well-designed system of transaction fees will likely lead to developers naturally learning most of the important points through personal experience.

Hence, there is substantial hope for a future that can be, to a significant degree, more decentralized; however, the days of easy gains are over. Now is the time for a much harder, and longer, slog of looking into the real world, and seeing how the technologies that we have built can actually benefit the world. During this stage, we will likely discover that at some point we will hit an inflection point, where most instances of "blockchain for x" will be made *not* by blockchain enthusiasts looking for something useful to do, coming upon x, and trying to do it, but rather by *enthusiasts of x* who look at blockchains and realize that they are a fairly useful tool for doing some part of x. Whether x is internet of things, financial infrastructure for the developing world, bottom-up social, cultural, and economic institutions, better data aggregation and protection for healthcare, or simply controversial charities and uncensorable marketplaces. In the latter two cases, the inflection point has likely already hit; many of the original crowd of blockchain enthusiasts *became* blockchain enthusiasts because of the politics. Once it hits in the other cases, however, then we will truly know that it has gone mainstream, and that the largest humanitarian gains are soon to come.

Additionally, we will likely discover that the concept of the "blockchain community" will cease to be meaningful as any kind of quasi-political movement in its own right; if any label applies at all, "crypto 2.0" is likely to be the most defensible one. The reason is similar to why we do not have a concept of the "distributed hash table community," and the "database community," while existent, is really simply a set of computer scientists who happen to specialize in databases: blockchains are just one technology, and so ultimately the greatest progress can only be achieved by working in combination with a whole other set of decentralized (and decentralization-friendly) technologies: reputation systems, distributed hash tables, "peer-to-peer hypermedia platforms," distributed messaging protocols, prediction markets, zero-knowledge proofs, and likely many more that have not yet been discovered.

PART 2: PROOF OF WORK

The Ethereum "genesis block" appeared on July 30, 2015, marking the beginning of the protocol's public life. This life did not find its footing easily. As the value of ETH tokens swelled to hundreds of millions of dollars, hackers attempted to exploit the system, requiring coordinated action from the burgeoning Ethereum community. Code alone, it turned out, was not enough to keep the system secure; human politics had a role too, and Buterin found himself at the center.

The most important of these trials was the hack of The DAO, an experimental collective venture fund that raised $150 million worth of ETH. ("DAO," pronounced like the first syllable of "Daoism," stands for "decentralized autonomous organization"—an organization built out of software on a blockchain.) Before it could begin investing, in June 2016, a hacker used a glitch in The DAO's code to withdraw funds from it. The DAO held what amounted to about 15% of the entire token supply, and handing a single user that kind of market share would prove especially dangerous if Ethereum made the transition to proof of stake, as Buterin intended. Counter-hackers delayed the hacker with countermeasures, while debates raged about whether the code must stand as is, glitches and all, or if something more drastic was necessary. Buterin championed the cause of the "hard fork"—an outright rewriting of the Ethereum blockchain to erase the hack. Although he held little formal power over the Ethereum protocol, the trust he had accrued proved decisive. Most of the Ethereum community followed him in placing the culture and the mission over the dictates of code.

Anxieties over his own charismatic authority are sprinkled between the lines of Buterin's writings at the time. Several months before the hack, on the Ethereum blog, he articulated a goal "to establish Ethereum as a decentralized project which is ultimately owned by all of humanity." During The DAO controversy, in "Why Cryptoeconomics and X-Risk Researchers Should Listen to Each Other More," he refers to an eventual "world-democracy DAO," perhaps a kind of United Nations based on direct participation. In "Control as Liability" he seems to compare himself to that other teenage founder of a globe-spanning network, Mark Zuckerberg; in the world of blockchains, in contrast to corporate platforms, central authority is better avoided than possessed. "On Free Speech" explores how the technology could actually prevent him from having the censorship powers to which Facebook and its ilk have increasingly succumbed. A 2018 tweet contends, "I think Ethereum can absolutely survive me spontaneously combusting tomorrow at this point." Yet the fact that he would have to say this at all suggests it might be short of a sure thing.

The year 2017 saw a seismic spike of value and traction for Ethereum. This was due largely to its use for "initial coin offerings," wherein startups (and far more outright scams) raised vast sums selling unregulated tokens on the promises in their whitepapers. Buterin publicly questioned the accuracy of Ethereum's market valuation and on Twitter urged the community "to differentiate between getting hundreds of billions of dollars of digital paper wealth sloshing around and actually achieving something meaningful for society." Ethereum was supposed to change the world, but as the examples in these essays suggest, a lot of the concrete use-cases were for things like financial games and gambling.

His writings during Ethereum's early years, rather than reveling in the price hikes and blockbuster token sales, dwelt in the design *problems of cryptoeconomics: How can incentives enable better kinds of coordination? Hard problems around identity and governance fascinate him here and in the more technical posts he was writing at the time. But as in the "Christmas Special" at the end of 2019, he also made time for play. Watching the intensity with which he and other Ethereans play chess at meetups, one can begin to wonder whether this whole multibillion-dollar experiment is really just a giant puzzle, a way of occupying the computing cycles coursing through their minds.*

—NS

WHY CRYPTOECONOMICS AND X-RISK RESEARCHERS SHOULD LISTEN TO EACH OTHER MORE

medium.com/@VitalikButerin
July 4, 2016

There has recently been a small but growing number of signs of interest in blockchains and cryptoeconomic systems from a community that has traditionally associated itself with artificial intelligence and various forms of futuristic existential risk research. Ralph Merkle, inventor of the now famous cryptographic technology which underpins Ethereum's light-client protocol, has expressed interest in DAO governance. Skype co-founder Jaan Tallinn proposed researching blockchain technology as a way to create mechanisms to solve global coordination problems. Prediction market advocates, who have long understood the potential of prediction markets as governance mechanisms, are now looking at Augur.* Is there anything interesting here? Is this simply a situation of computer geeks who were previously attracted to computer-geek-friendly topic A now also being attracted to a

---

* Augur is a crypto prediction market platform that allows users to bet on particular events, together with an "oracle" system that determines the real-world outcomes of those events.

completely unrelated but also computer-geek-friendly topic B, or is there an actual connection?

I would argue that there is, and the connection is as follows. **Both the cryptoeconomics research community and the AI safety / new cyber-governance / existential risk community are trying to tackle what is fundamentally the same problem: How can we regulate a very complex and very smart system with unpredictable emergent properties using a very simple and dumb system whose properties once created are inflexible?**

In the context of AI research, a major sub-problem is that of defining a utility function that would guide the behavior of a superintelligent agent without accidentally guiding it into doing something that satisfies the function as written but does not satisfy the intent (sometimes called "edge instantiation"). For example, if you tried to tell a superintelligent AI to cure cancer, it may end up reasoning that the most reliable way to do that is to simply kill everyone first. If you tried to plug that hole, it may decide to permanently cryogenically freeze all humans without killing them. And so forth. In the context of Ralph Merkle's DAO democracy, the problem is that of determining an objective function that is correlated with social and technological progress and generally things that people want, is anti-correlated with existential risks, and is easily measurable enough that its measurement would not itself become a source of political battles.

Meanwhile, in the context of cryptoeconomics, the problems are surprisingly similar. The core problem of consensus asks how to incentivize validators to continue supporting and growing a coherent history using a simple algorithm that is set in stone, when the validators themselves are highly complex economic agents that are free to interact in arbitrary ways. The issue found with The DAO was a divergence of software developers' complex intent, having a specific use in mind for the splitting function, and the de facto result of the software implementation. Augur tries to extend the consensus problem to real-world facts. Maker is trying to create a decentralized-governance algorithm for a platform that intends to provide an asset with the decentralization of cryptocurrency and the reliability of fiat. In all of these cases, the algorithms are dumb, and yet the agents that they have to control are quite smart. AI safety is about agents with IQ 150 trying to control agents with IQ 6,000, whereas cryptoeconomics is about agents with IQ 5 trying to control agents with IQ 150—problems that are certainly different, but the similarities are not to be scoffed at.

These are all hard problems, and they are problems that both communities have already been separately considering for many years and have in some cases amassed considerable insights about. They are also problems where heuristic partial solutions and mitigation strategies are already starting to be discovered. In the case of DAOs, some developers are moving toward a hybrid approach that has a set of curators with some control over the DAO's assets, but assigns those curators only limited powers that are by themselves enough to rescue a DAO from an attack, but not enough to unilaterally carry out an attack that causes more than moderate disruption—an approach with some similarities to ongoing research into safe AI interruptibility.

On the futarchy side, people are looking at interest rates as an objective function, a kind of hybrid of futarchy and quadratic voting* through voluntary coin locking as a governance algorithm, and various forms of moderated futarchy that give the futarchy enough power to prevent a majority collusion attack in a way that

---

* Quadratic voting is a mechanism in which voters can vote with multiple tokens, but the more tokens one votes with, the less power each token has. It is a system that seeks to account for intensity of preference while counteracting a plutocracy by those who simply hold the most tokens.

a democracy cannot, but otherwise leave the power to a voting process—all innovations that are at least worth the consideration of a group trying to use futarchy to build a world-democracy DAO.

Another highly underappreciated solution is the use of governance algorithms that explicitly slow things down—the proposed DAO hard fork that may rescue the contained funds is only possible precisely because The DAO included a set of rules that required every action to have a long delay time. Still another avenue that is starting to be explored is formal verification—using computer programs to automatically verify other computer programs, and make sure that they satisfy a set of claims about what the programs are supposed to do.

Formally proving "honesty" in the general case is impossible, due to the complexity of value problem, but we can make some partial guarantees to reduce risk. For example, we could formally prove that a certain kind of action cannot be taken in less than seven days, or that a certain kind of action cannot be taken for forty-eight hours, if the curators of a given DAO vote to flip a switch. In an AI context, such proofs could be used to prevent certain kinds of simple bugs in the reward function that would result in a completely unintended behavior appearing to the AI to be of extremely high value. Of course, many other communities have been thinking about formal verification for many years already, but now it is being explored for a different use in a novel setting.

Meanwhile, one example of a concept promoted in the AI safety circles that may be highly useful to those building economic systems containing DAOs is superrational-decision theories—essentially, ways to overcome prisoner's-dilemma situations by committing to run source code that treats agents which also commit to run that source code more favorably. One example of a move available to open-source agents that is not available to "black box" agents is the "values handshake" described in a short story by Scott Alexander: two agents can agree to both commit to maximize a goal which is the average of the two goals that they previously had. Previously, such concepts were largely science fiction, but now futarchy DAOs can actually do this. More generally, a DAO may well be a highly effective means for a social institution to strongly commit to "running source code" that has particular properties.

"The DAO" is only the first in a series of many that will be launched over the course of this year and the next, and you can bet that all of the subsequent examples will learn heavily from the lessons of the first one, and each will come up with different and innovative software-code security policies, governance algorithms, curator systems, slow and phased bootstrap and rollout processes, and formally verified guarantees in order to do its best to make sure that it can weather the cryptoeconomic storm.

Finally, I would argue that the biggest lesson to learn from the crypto community is that of decentralization itself: have different teams implement different pieces redundantly, so as to minimize the chance that an oversight from one system will pass through the other systems undetected. The crypto ecosystem is shaping up to be a live experiment comprising many challenges at the forefront of software development, computer science, game theory, and philosophy, and the results, regardless of whether they make it into mainstream social applications in their present form or after several iterations that involve substantial changes to the core concepts, are welcome for anyone to learn from and see.

A PROOF-OF-STAKE DESIGN PHILOSOPHY

medium.com/@VitalikButerin
December 30, 2016

Systems like Ethereum (and Bitcoin, and NXT, and BitShares, etc.) are a fundamentally new class of cryptoeconomic organisms—decentralized, jurisdiction-less entities that exist entirely in cyberspace, maintained by a combination of cryptography, economics, and social consensus. They are kind of like BitTorrent, but they are also not like BitTorrent, as BitTorrent has no concept of state—a distinction that turns out to be crucially important. They are sometimes described as decentralized autonomous corporations, but they are also not quite corporations—you can't hard fork Microsoft. They are kind of like open-source software projects, but they are not quite that either—you can fork a blockchain, but not quite as easily as you can fork OpenOffice.*

These cryptoeconomic networks come in many flavors—ASIC-based PoW, GPU-based PoW, naïve PoS, delegated PoS,

---

* OpenOffice is a free, open-source office suite similar to Microsoft Office. To "fork" open-source software means to copy its freely available code and modify it into something different.

hopefully soon Casper PoS*—and each of these flavors inevitably comes with its own underlying philosophy. One well-known example is the maximalist vision of proof of work, where "the" correct blockchain, singular, is defined as the chain that miners have burned the largest amount of economic capital to create. Originally a mere in-protocol fork choice rule, this mechanism has in many cases been elevated to a sacred tenet. BitShares' delegated proof of stake presents another coherent philosophy, where everything once again flows from a single tenet, but one that can be described even more simply: shareholders vote.

Each of these philosophies—Nakamoto consensus, social consensus, shareholder voting consensus—leads to its own set of conclusions and to a system of values that makes quite a bit of sense when viewed on its own terms—though they can certainly be criticized when compared against each other. Casper consensus has a philosophical underpinning too, though one that has so far not been as succinctly articulated.

Myself, Vlad, Dominic, Jae, and others all have their own views on why proof-of-stake protocols exist and how to design them, but here I intend to explain where I personally am coming from.

I'll proceed to listing observations and then conclusions directly:

□ Cryptography is truly special in the twenty-first century because **cryptography is one of the very few fields where adversarial conflict continues to heavily favor the defender**. Castles are far easier to destroy than build, islands are defendable but can still be attacked, but an average person's ECC keys are secure enough to resist even state-level actors.

---

* Casper PoS is the algorithm designed to support Ethereum's conversion to proof of stake, using a betting system to prevent malicious actors.

Cypherpunk philosophy is fundamentally about leveraging this precious asymmetry to create a world that better preserves the autonomy of the individual, and cryptoeconomics is to some extent an extension of that, except this time protecting the safety and liveness of complex systems of coordination and collaboration, rather than simply the integrity and confidentiality of private messages. **Systems that consider themselves ideological heirs to the cypherpunk spirit should maintain this basic property, and be much more expensive to destroy or disrupt than they are to use and maintain.**

□ The "cypherpunk spirit" isn't just about idealism; making systems that are easier to defend than they are to attack is also simply sound engineering.

□ **On medium-to-long time scales, humans are quite good at consensus.** Even if an adversary had access to unlimited hashing power, and came out with a 51% attack of any major blockchain that reverted even the last month of history, convincing the community that this chain is legitimate is much harder than just outrunning the main chain's hashpower. They would need to subvert block explorers, every trusted member in the community, the *New York Times*, archive.org, and many other sources on the internet; all in all, convincing the world that the new attack chain is the one that came first in the information-technology-dense twenty-first century is about as hard as convincing the world that the US moon landings never happened. **These social considerations are what ultimately protect any blockchain in the long term**, regardless of whether or not the blockchain's community admits it (note that Bitcoin Core does admit this primacy of the social layer).

☐ However, a blockchain protected by social consensus alone would be far too inefficient and slow, and it would be too easy for disagreements to continue without end (though despite all difficulties, it has happened); hence, economic consensus serves an extremely important role in protecting liveness and safety properties in the short term.

☐ Because proof-of-work security can only come from block rewards, and incentives to miners can only come from the risk of them losing their future block rewards, **proof of work necessarily operates on a logic of massive power incentivized into existence by massive rewards**. Recovery from attacks in PoW is very hard: the first time it happens, you can hard fork to change the PoW and thereby render the attacker's ASICs useless, but the second time you no longer have that option, and so the attacker can attack again and again. Hence, the size of the mining network has to be so large that attacks are inconceivable. Attackers of size less than x are discouraged from appearing by having the network constantly spend x every single day. **I reject this logic because (i) it kills trees, and (ii) it fails to realize the cypherpunk spirit—cost of attack and cost of defense are at a one-to-one ratio, so there is no defender's advantage.**

☐ **Proof of stake breaks this symmetry by relying not on rewards for security, but rather on penalties.** Validators put money ("deposits") at stake, are rewarded slightly to compensate them for locking up their capital and maintaining nodes and taking extra precaution to ensure their private key safety, but the bulk of the cost of reverting transactions comes from penalties that are hundreds or thousands of times larger than the rewards that they got in the meantime. **The "one-sentence philosophy" of proof of stake is thus not "security comes from burning energy," but rather "security comes from putting up economic value-at-loss."** A given block or state has x-dollar security if you can prove that achieving an equal level of finalization for any conflicting block or state cannot be accomplished unless malicious nodes complicit in an attempt to make the switch pay x dollars' worth of in-protocol penalties.

□ Theoretically, a majority collusion of validators may take over a proof-of-stake chain and start acting maliciously. However, (i) through clever protocol design, their ability to earn extra profits through such manipulation can be limited as much as possible; and more importantly, (ii) if they try to prevent new validators from joining, or execute 51% attacks, then the community can simply coordinate a hard fork and delete the offending validators' deposits. **A successful attack may cost $50 million, but the process of cleaning up the consequences will not be that much more onerous than the Geth-Parity consensus failure of November 25, 2016.*** Two days later, the blockchain and community are back on track, attackers are $50 million poorer, and the rest of the community is likely richer since the attack will have caused the value of the token to go *up* due to the ensuing supply crunch. *That's* attack-defense asymmetry for you.

□ The above should not be taken to mean that unscheduled hard forks will become a regular occurrence; if desired, the cost of a *single* 51% attack on proof of stake can certainly

---

* This refers to a bug in the popular Go Ethereum client that required a rapid software update that briefly forked the blockchain, with two different concurrent ledgers.

be set to be as high as the cost of a *permanent* 51% attack on proof of work, and the sheer cost and ineffectiveness of an attack should ensure that it is almost never attempted in practice.

☐ **Economics is not everything.** Individual actors may be motivated by extra-protocol motives, they may get hacked, they may get kidnapped, or they may simply get drunk and decide to wreck the blockchain one day and to hell with the cost. Furthermore, on the bright side, **individuals' moral forbearances and communication inefficiencies will often raise the cost of an attack to levels much higher than the nominal protocol-defined value-at-loss**. This is an advantage that we cannot rely on, but at the same time it is an advantage that we should not needlessly throw away.

☐ **Hence, the best protocols are protocols that work well under a variety of models and assumptions**—economic rationality with coordinated choice, economic rationality with individual choice, simple fault tolerance, Byzantine fault tolerance (ideally both the adaptive and non-adaptive adversary variants), Ariely- and Kahneman-inspired behavioral economic models ("we all cheat just a little"), and ideally any other model that's realistic and practical to reason about. **It is important to have both layers of defense: economic incentives to discourage centralized cartels from acting antisocially, and anti-centralization incentives to discourage cartels from forming in the first place.**

☐ **Consensus protocols that work as-fast-as-possible have risks and should be approached very carefully if at all**, because if the *possibility* to be very fast is tied to *incentives* to do so, the combination will reward very high and systemic-risk-inducing levels of **network-level centralization** (e.g., all validators running from the same hosting provider). Consensus protocols that don't care too much how fast a validator sends a message, as long as they do so within some acceptably long time interval (e.g., four to eight seconds, as we empirically know that latency in Ethereum is usually around five hundred milliseconds to one second). A possible middle ground is to create protocols that can work very quickly, but where mechanics similar to Ethereum's uncle mechanism* ensure that the marginal reward for a node increasing its degree of network connectivity beyond some easily attainable point is fairly low.

From here, there are of course many details and many ways to diverge on the details, but the above are the core principles that at least my version of Casper is based on. From here, we can certainly debate tradeoffs between competing values. Do we give ETH a 1% annual issuance rate and get a $50 million cost of forcing a remedial hard fork, or a zero annual issuance rate and get a $5 million cost of forcing a remedial hard fork? When do we increase a protocol's security under the economic model in exchange for decreasing its security under a fault-tolerance model? Do we care more about having a predictable level of security or a predictable level of issuance? These are all questions for another post, and the various ways of *implementing* the different tradeoffs between these values are questions for yet more posts. But we'll get to it :)

---

* In Ethereum, "uncle blocks" are incomplete blocks not ultimately added to the main chain. Miners receive a reward for producing these, however—a kind of consolation prize for how their failed efforts contribute to the security of the system as a whole.

THE MEANING OF DECENTRALIZATION

medium.com/@VitalikButerin
February 6, 2017

"Decentralization" is one of the most frequently used words in the cryptoeconomics space and is often even viewed as a blockchain's entire raison d'être, but it is also perhaps one of the most poorly defined words. Thousands of hours of research, and billions of dollars of hashpower, have been spent for the sole purpose of attempting to achieve decentralization, and to protect and improve it, and when discussions get rivalrous it is extremely common for proponents of one protocol (or protocol extension) to claim, as the ultimate knockdown argument, that the opposing proposals are "centralized."

But there is often a lot of confusion as to what this word actually means. Consider, for example, the following completely unhelpful, but unfortunately all-too-common, diagram:*

---

* This diagram comes from Paul Baran, *On Distributed Communications* (RAND Corporation, 1964), a memo that proposed the network structure for what would become the internet.

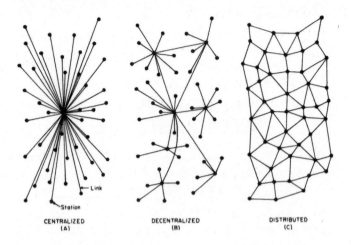

CENTRALIZED   DECENTRALIZED   DISTRIBUTED
(A)   (B)   (C)

Now, consider the two answers on Quora for "What is the difference between distributed and decentralized?" The first essentially parrots the above diagram, whereas the second makes the entirely different claim that "distributed means not all the processing of the transactions is done in the same place," whereas "decentralized means that not one single entity has control over all the processing." Meanwhile, the top answer on the Ethereum Stack Exchange gives a very similar diagram, but with the words "decentralized" and "distributed" having switched places! Clearly, a clarification is in order.

THREE TYPES OF DECENTRALIZATION

When people talk about software decentralization, there are actually *three separate axes* of centralization/decentralization that they may be talking about. While in some cases it is difficult to see how you can have one without the other, in general they are quite independent of each other. The axes are as follows:

- **ARCHITECTURAL (DE)CENTRALIZATION:** How many **physical computers** is a system made up of? How many of those computers can it tolerate breaking down at any single time?

- **POLITICAL (DE)CENTRALIZATION:** How many **individuals or organizations** ultimately control the computers that the system is made up of?

- **LOGICAL (DE)CENTRALIZATION:** Does the **interface and data structures** that the system presents and maintains look more like a single monolithic object, or an amorphous swarm? One simple heuristic is: If you cut the system in half, including both providers and users, will both halves continue to fully operate as independent units?

We can try to put these three dimensions into a chart:

	Logically centralized		Logically decentralized		
	Politically centralized	Politically decentralized	Politically centralized	Politically decentralized	
Architecturally centralized	Traditional corporations	Direct democracy	?	?	Architecturally centralized
	Civil law				
Architecturally decentralized	?	Blockchains, Common law	Traditional CDNs, Esperanto (initially)	BitTorrent, English language	Architecturally decentralized

Note that a lot of these placements are very rough and highly debatable. But let's try going through any of them:

- Traditional corporations are politically centralized (one CEO), architecturally centralized (one head office), and logically centralized (can't really split them in half).

☐ Civil law relies on a centralized law-making body, whereas common law is built up of precedent made by many individual judges. Civil law still has some architectural decentralization as there are many courts that nevertheless have large discretion, but common law has more of it. Both are logically centralized ("the law is the law").

☐ Languages are logically decentralized; the English spoken between Alice and Bob and the English spoken between Charlie and David do not need to agree at all. There is no centralized infrastructure required for a language to exist, and the rules of English grammar are not created or controlled by any one single person (whereas Esperanto was originally invented by Ludwik Zamenhof, though now it functions more like a living language that evolves incrementally with no authority).

☐ BitTorrent is logically decentralized similarly to how English is. Content-delivery networks are similar, but are controlled by one single company.

☐ Blockchains are politically decentralized (no one controls them) and architecturally decentralized (no infrastructural central point of failure) but they are logically centralized (there is one commonly agreed state and the system *behaves* like a single computer).

Many times when people talk about the virtues of a blockchain, they describe the convenience benefits of having "one central database"; that centralization is logical centralization, and it's a kind of centralization that is arguably in many cases good (though Juan Benet from IPFS would also push for logical decentralization wherever possible, because logically decentralized systems tend to be good at surviving network partitions, work well in regions of the world that have poor connectivity, etc.).

Architectural centralization often leads to political centralization, though not necessarily—in a formal democracy, politicians meet and hold votes in some physical governance chamber, but the maintainers of this chamber do not end up deriving any substantial amount of power over decision-making as a result. In computerized systems, architectural but not political decentralization might happen if there is an online community which uses a centralized forum for convenience, but where there is a widely agreed social contract that if the owners of the forum act maliciously then everyone will move to a different forum (communities that are formed around rebellion against what they see as censorship in another forum likely have this property in practice).

Logical centralization makes architectural decentralization harder, but not impossible—see how decentralized consensus networks have already been proven to work, but are more difficult than maintaining BitTorrent. And logical centralization makes political decentralization harder—in logically centralized systems, it's harder to resolve contention by simply agreeing to "live and let live."

THREE REASONS FOR DECENTRALIZATION

The next question is: Why is decentralization useful in the first place? There are generally several arguments raised:

□ **FAULT TOLERANCE:** Decentralized systems are less likely to fail accidentally because they rely on many separate components that are not likely.

□ **ATTACK RESISTANCE:** Decentralized systems are more expensive to attack and destroy or manipulate because they lack sensitive central points that can be attacked at much lower cost than the economic size of the surrounding system.

☐ **COLLUSION RESISTANCE:** It is much harder for participants in decentralized systems to collude to act in ways that benefit them at the expense of other participants, whereas the leaderships of corporations and governments collude in ways that benefit themselves but harm less well-coordinated citizens, customers, employees, and the general public all the time.

All three arguments are important and valid, but all three arguments lead to some interesting and different conclusions once you start thinking about protocol decisions with the three individual perspectives in mind. Let us try to expand out each of these arguments one by one.

Regarding fault tolerance, the core argument is simple. What's less likely to happen: one single computer failing or five out of ten computers all failing at the same time? The principle is uncontroversial, and is used in real life in many situations, including jet engines, backup power generators (particularly in places like hospitals), military infrastructure, financial portfolio diversification, and yes, computer networks.

However, this kind of decentralization, while still effective and highly important, often turns out to be far less of a panacea than a naïve mathematical model would sometimes predict. The reason is common mode failure. Sure, four jet engines are less likely to fail than one jet engine, but what if all four engines were made in the same factory, and a fault was introduced in all four by the same rogue employee?

Do blockchains as they are today manage to protect against common mode failure? Not necessarily. Consider the following scenarios:

- ☐ All nodes in a blockchain run the same client software, and this client software turns out to have a bug.

- ☐ All nodes in a blockchain run the same client software, and the development team of this software turns out to be socially corrupted.

- ☐ The research team that is proposing protocol upgrades turns out to be socially corrupted.

- ☐ In a proof-of-work blockchain, 70% of miners are in the same country, and the government of this country decides to seize all mining farms for national-security purposes.

- ☐ The majority of mining hardware is built by the same company, and this company gets bribed or coerced into implementing a backdoor that allows this hardware to be shut down at will.

- ☐ In a proof-of-stake blockchain, 70% of the coins at stake are held at one exchange.

A holistic view of fault-tolerance decentralization would look at all of these aspects, and see how they can be minimized. Some natural conclusions that arise are fairly obvious:

- ☐ It is crucially important to have multiple competing implementations.

- ☐ The knowledge of the technical considerations behind protocol upgrades must be democratized, so that more people can feel comfortable participating in research discussions and criticizing protocol changes that are clearly bad.

- ☐ Core developers and researchers should be employed by

multiple companies or organizations (or, alternatively, many of them can be volunteers).

□ Mining algorithms should be designed in a way that minimizes the risk of centralization.

□ Ideally we use proof of stake to move away from hardware-centralization risk entirely (though we should also be cautious of new risks that pop up due to proof of stake).

Note that the fault-tolerance requirement in its naïve form focuses on architectural decentralization, but once you start thinking about fault tolerance of the community that governs the protocol's ongoing development, then political decentralization is important too.

Now, let's look at attack resistance. In some pure economic models, you sometimes get the result that decentralization does not even matter. If you create a protocol where the validators are guaranteed to lose $50 million if a 51% attack (i.e., finality reversion) happens, then it doesn't really matter if the validators are controlled by one company or one hundred companies—$50 million economic security margin is $50 million economic security margin. In fact, there are deep game-theoretic reasons why centralization may even *maximize* this notion of economic security (the transaction selection model of existing blockchains reflects this insight, as transaction inclusion into blocks through miners and block proposers is actually a very rapidly rotating dictatorship).

However, once you adopt a richer economic model, and particularly one that admits the possibility of coercion (or much milder things like targeted DoS attacks against nodes), decentralization becomes more important. If you threaten one person with death, suddenly $50 million will not matter to them as much anymore. But if the $50 million is spread between ten people, then you have to threaten ten times as many people, and do it all at the same

time. In general, the modern world is in many cases characterized by an attack-defense asymmetry in favor of the attacker—a
building that costs $10 million to build may cost less than
$100,000 to destroy, but the attacker's leverage is often sublinear:
if a building that costs $10 million to build costs $100,000 to
destroy, a building that costs $1 million to build may realistically
cost perhaps $30,000 to destroy. Smaller gives better ratios.

What does this reasoning lead to? First of all, it pushes strongly in
favor of proof of stake over proof of work, as computer hardware is
easy to detect, regulate, or attack, whereas coins can be much more
easily hidden (proof of stake also has strong attack resistance for
other reasons). Second, it is a point in favor of having widely distributed development teams, including geographic distribution. Third,
it implies that both the economic model and the fault-tolerance
model need to be looked at when designing consensus protocols.

Finally, we can get to perhaps the most intricate argument of
the three, collusion resistance. Collusion is difficult to define;
perhaps the only truly valid way to put it is to simply say that
collusion is "coordination that we don't like." There are many situations in real life where even though having perfect coordination
between everyone would be ideal, one subgroup being able to
coordinate *while the others cannot* is dangerous.

One simple response is antitrust law—deliberate regulatory
barriers that get placed in order to make it more difficult for participants on one side of the marketplace to come together and
act like a monopolist and get outsized profits at the expense of
both the other side of the marketplace and general social welfare.
Another example is rules against active coordination between candidates and super PACs in the United States, though those have
proven difficult to enforce in practice. A much smaller example
is a rule in some chess tournaments preventing two players from
playing many games against each other to try to raise one player's

score. No matter where you look, attempts to prevent undesired coordination in sophisticated institutions are everywhere.

In the case of blockchain protocols, the mathematical and economic reasoning behind the safety of the consensus often relies crucially on the uncoordinated-choice model, or the assumption that the game consists of many small actors that make decisions independently. If any one actor gets more than one-third of the mining power in a proof-of-work system, they can gain outsized profits by selfish-mining.* However, can we really say that the uncoordinated-choice model is realistic when 90% of the Bitcoin network's mining power is well-coordinated enough to show up together at the same conference?

Blockchain advocates also make the point that blockchains are more secure to build on because they can't just change their rules arbitrarily on a whim whenever they want to, but this case would be difficult to defend if the developers of the software and protocol were all working for one company, were part of one family, and sat in one room. *The whole point* is that these systems should

---

* This is a strategy in which miners might collude to produce a private chain and corrupt the validity of the main chain.

not act like self-interested unitary monopolies. Hence, you can certainly make a case that blockchains would be more secure if they were more *discoordinated*.

However, this presents a fundamental paradox. Many communities, including Ethereum's, are often praised for having a strong community spirit and being able to coordinate quickly on implementing, releasing, and activating a hard fork to fix denial-of-service issues in the protocol within six days. But how can we foster and improve this good kind of coordination, but at the same time prevent "bad coordination" that consists of miners trying to screw everyone else over by repeatedly coordinating 51% attacks?

There are three ways to answer this:

□ Don't bother mitigating undesired coordination; instead, try to build protocols that can resist it.

□ Try to find a happy medium that allows enough coordination for a protocol to evolve and move forward, but not enough to enable attacks.

□ Try to make a distinction between beneficial coordination and harmful coordination, and make the former easier and the latter harder.

The first approach makes up a large part of the Casper design philosophy. However, it by itself is insufficient, as relying on economics alone fails to deal with the other two categories of concerns about decentralization. The second is difficult to engineer explicitly, especially for the long term, but it does often happen accidentally. For example, the fact that Bitcoin's core developers generally speak English but miners generally speak Chinese can be viewed as a happy accident, as it creates a kind of "bicameral" governance that makes coordination more difficult, with the side benefit of reducing the risk of common mode failure, as the English and Chinese communities will reason at least somewhat separately due to distance and communication difficulties and are therefore less likely to both make the same mistake.

The third is a social challenge more than anything else; solutions in this regard may include:

□ Social interventions that try to increase participants' loyalty to the community around the blockchain as a whole and substitute or discourage the possibility of the players on one side of a market becoming directly loyal to each other.

□ Promoting communication between different "sides of the market" in the same context, so as to reduce the possibility that validators or developers or miners begin to see themselves as a "class" that must coordinate to defend their interests against other classes.

□ Designing the protocol in such a way as to reduce the incentive for validators and miners to engage in one-to-one "special relationships," centralized relay networks, and other similar super-protocol mechanisms.

□ Clear norms about what are the fundamental properties that the protocol is supposed to have, and what kinds of things should not be done, or at least should be done only under very extreme circumstances.

This third kind of decentralization, decentralization as undesired-coordination avoidance, is thus perhaps the most difficult to achieve, and tradeoffs are unavoidable. Perhaps the best solution may be to rely heavily on the one group that is guaranteed to be fairly decentralized: the protocol's users.

NOTES ON BLOCKCHAIN GOVERNANCE

vitalik.ca
December 17, 2017

One of the more interesting recent trends in blockchain governance is the resurgence of on-chain coin-holder voting as a multipurpose decision mechanism. Votes by coin holders are sometimes used in order to decide who operates the super-nodes that run a network (e.g., DPoS in EOS, NEO, Lisk, and other systems), sometimes to vote on protocol parameters (e.g., the Ethereum gas limit) and sometimes to vote on and directly implement protocol upgrades wholesale (e.g., Tezos). In all of these cases, the votes are automatic—the protocol itself contains all of the logic needed to change the validator set or to update its own rules, and does this automatically in response to the result of votes.

Explicit on-chain governance is typically touted as having several major advantages. First, unlike the highly conservative philosophy espoused by Bitcoin, it can evolve rapidly and accept needed technical improvements. Second, by creating an explicit decentralized framework, it avoids the perceived pitfalls of *informal* governance, which is viewed to be either too unstable and prone to chain splits, or prone to becoming too de facto centralized—the latter being the same argument made in the famous 1972 essay "Tyranny of Structurelessness."*

Quoting Tezos documentation:

> While all blockchains offer financial incentives for maintaining consensus on their ledgers, no blockchain has a robust on-chain mechanism that seamlessly amends the rules governing its protocol and rewards protocol development. As a result, first-generation blockchains empower de facto, centralized core development teams or miners to formulate design choices.

And:

> Yes, but why would you want to make [a minority chain split] easier? Splits destroy network effects.

On-chain governance used to select validators also has the benefit that it allows for networks that impose high computational performance requirements on validators without introducing economic centralization risks and other traps of the kind that appear in public blockchains.

So far, all in all, on-chain governance seems like a very good bargain . . . so what's wrong with it?

WHAT IS BLOCKCHAIN GOVERNANCE?

To start off, we need to describe more clearly what the process of "blockchain governance" *is*. Generally speaking, there are two

---

* By Jo Freeman—a reflection on the informal hierarchies that arose in allegedly nonhierarchical feminist "rap groups," and an analysis that has been frequently applied to the informal hierarchies that arise in online communities.

informal models of governance, that I will call the "decision function" view of governance and the "coordination" view of governance. The decision function view treats governance as a function $f(x_1, x_2 \ldots x_n) \rightarrow y$, where the inputs are the wishes of various legitimate stakeholders (senators, the president, property owners, shareholders, voters, etc.) and the output is the decision.

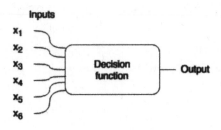

The decision function view is often useful as an approximation, but it clearly frays very easily around the edges: people often can and do break the law and get away with it, sometimes rules are ambiguous, and sometimes revolutions happen—and all three of these possibilities are, at least sometimes, *a good thing*. And often even behavior inside the system is shaped by incentives created by *the possibility* of acting outside the system, and this once again is at least sometimes a good thing.

The coordination model of governance, in contrast, sees governance as something that exists in layers. The bottom layer is, in the real world, the laws of physics themselves (as a geopolitical realist would say, guns and bombs), and in the blockchain space we can abstract a bit further and say that it is each individual's ability to run whatever software they want in their capacity as a user, miner, stakeholder, validator, or whatever other kind of agent a blockchain protocol allows them to be. The bottom layer is always the ultimate deciding layer; if, for example, all Bitcoin users wake up one day and decide to edit their clients' source code and replace the entire code with an Ethereum client that listens to balances of a particular ERC20 token contract, then that means that that ERC20 token *is* bitcoin. The bottom layer's ultimate governing power cannot be stopped, but the actions that people take on this layer can be *influenced* by the layers above it.

The second (and crucially important) layer is coordination institutions. The purpose of a coordination institution is to create focal points around how and when individuals should act in order to better coordinate behavior. There are many situations, both in blockchain governance and in real life, where if you act in a certain way alone, you are likely to get nowhere (or worse), but if everyone acts together a desired result can be achieved.

	A	B
A	(5, 5)	(0, 0)
B	(0, 0)	(5, 5)

An abstract coordination game. You benefit heavily from
making the same move as everyone else.

In these cases, it's in your interest to go if everyone else is going, and stop if everyone else is stopping. You can think of coordination institutions as putting up green or red flags in the air that say "go" or "stop," *with an established culture* in which everyone watches these flags and (usually) does what they say. Why do people have the incentive to follow these flags? Because *everyone else* is already following these flags, and you have the incentive to do the same thing as what everyone else is doing.

A Byzantine general* rallying his troops forward. The purpose of this isn't just to make the soldiers feel brave and excited, but also to reassure them that everyone else feels brave and excited and will charge forward as well, so an individual soldier is not just committing suicide by charging forward alone.

> **STRONG CLAIM:** This concept of coordination flags encompasses *all* that we mean by "governance"; in scenarios where coordination games (or more generally, multi-equilibrium games) do not exist, the concept of governance is meaningless.

In the real world, military orders from a general function as a flag, and in the blockchain world, the simplest example of such a flag is the mechanism that tells people whether or not a hard fork "is happening." Coordination institutions can be very formal, or they can be informal, and often give suggestions

---

* The use of this example is an ode to the Byzantine generals problem in game theory: A circle of armies surround Byzantium, and they all need to attack at the same time to win. If they lack a secure means of communicating with each other, how can they coordinate a simultaneous attack?

that are ambiguous. Flags would ideally always be either red or green, but sometimes a flag might be yellow, or even holographic, appearing green to some participants and yellow or red to others. Sometimes there are also multiple flags that conflict with each other.

The key questions of governance thus become:

☐ What should layer 1 be? That is, what features should be set up in the initial protocol itself, and how does this influence the ability to make formulaic (i.e., decision-function-like) protocol changes, as well as the level of power of different kinds of agents to act in different ways?

☐ What should layer 2 be? That is, what coordination institutions should people be encouraged to care about?

THE ROLE OF COIN VOTING

Ethereum also has a history with coin voting, including:

☐ **DAO PROPOSAL VOTES:** daostats.github.io/proposals.html

☐ **THE DAO CARBONVOTE:** v1.carbonvote.com

☐ **THE EIP 186/649/669 CARBONVOTE:** carbonvote.com

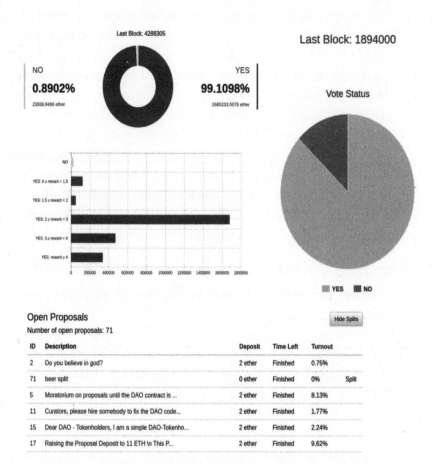

These three are all examples of *loosely coupled* coin voting, or coin voting as a layer 2 coordination institution. Ethereum does not have any examples of *tightly coupled* coin voting (or coin voting as a layer 1 in-protocol feature), though it *does* have an example of tightly coupled *miner* voting: miners' right to vote on the gas limit.* Clearly, tightly coupled voting and loosely coupled voting

---

* In the sense used here, the gas limit is the ceiling that miners collectively impose on how much network activity they will permit within a single block. It is a way of balancing the capacity of the system with the resource expenditure required of miners.

are competitors in the governance-mechanism space, so it's worth dissecting: What are the advantages and disadvantages of each one?

Assuming zero transaction costs, and if used as a sole governance mechanism, the two are clearly equivalent. If a loosely coupled vote says that change X should be implemented, then that will serve as a "green flag" encouraging everyone to download the update; if a minority wants to rebel, they will simply not download the update. If a tightly coupled vote implements change X, then the change happens automatically, and if a minority wants to rebel they can install a hard fork update that cancels the change. However, there clearly are nonzero transaction costs associated with making a hard fork, and this leads to some very important differences.

One very simple, and important, difference is that tightly coupled voting creates a default in favor of the blockchain adopting what the majority wants, requiring minorities to exert great effort to coordinate a hard fork to preserve a blockchain's existing properties, whereas loosely coupled voting is only a coordination tool, and still requires users to actually download and run the software that implements any given fork. But there are also many other differences. Now, let us go through some arguments *against* voting, and dissect how each argument applies to voting as layer 1 and voting as layer 2.

LOW VOTER PARTICIPATION

One of the main criticisms of coin-voting mechanisms so far is that, no matter where they are tried, they tend to have very low voter participation. The DAO Carbonvote only had a voter participation rate of 4.5%:

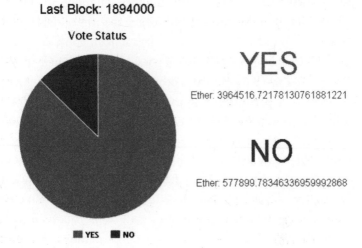

Additionally, wealth distribution is very unequal, and the results of these two factors together are best described by this image created by a critic of the DAO fork:

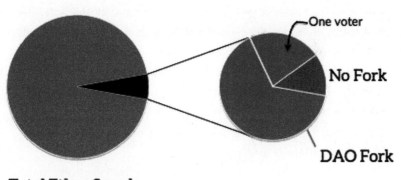

The EIP 186 Carbonvote had about 2.7 million ETH voting. The DAO proposal votes did not fare better, with participation never reaching 10%. And outside of Ethereum things are not sunny either; even in BitShares, a system where the core social contract is designed around voting, the top delegate in an approval vote only got 17% of the vote, and in Lisk it got up to 30%, though as we will discuss later these systems have other problems of their own.

Low voter participation means two things. First, the vote has a harder time achieving a perception of legitimacy, because it only reflects the views of a small percentage of people. Second, an attacker with only a small percentage of all coins can sway the vote. These problems exist regardless of whether the vote is tightly coupled or loosely coupled.

GAME-THEORETIC ATTACKS

Aside from "the big hack" that received the bulk of the media attention, The DAO also had a number of much smaller game-theoretic vulnerabilities. But this is only the tip of the iceberg. Even if all of the finer details of a voting mechanism are implemented correctly, voting mechanisms in general have a large flaw: in any vote, the probability that any given voter will have an impact on the result is tiny, and so the personal incentive that each voter has to vote honestly is almost insignificant. And if each person's size of the stake is small, their incentive to vote correctly is insignificant *squared*. Hence, a relatively small bribe spread out across the participants may suffice to sway their decision, possibly in a way that they collectively might quite disapprove of.

Now you might say, people are not evil, selfish profit maximizers that will accept a $0.50 bribe to vote to give $20 million to Josh Garza* just because the above calculation says their individual

---

* The CEO of the crypto mining company GAW Miners, Josh Garza plead guilty to wire fraud and was sentenced to prison time in 2018 for running a Ponzi scheme.

chance of affecting anything is tiny; rather, they would altruistically refuse to do something that evil. There are two responses to this criticism.

First, there are ways to make a "bribe" that are quite plausible; for example, an exchange can offer interest rates for deposits (or, even more ambiguous, use the exchange's own money to build a great interface and features), with the exchange operator using the large quantity of deposits to vote as they wish. Exchanges profit from chaos, so their incentives are clearly quite misaligned with users *and* coin holders.

Second, and more damningly, in practice it seems like people, at least in their capacity as crypto-token holders, are profit maximizers, and seem to see nothing evil or selfish about taking a bribe or two. As "Exhibit A," we can look at the situation with Lisk, where the delegate pool seems to have been successfully captured by two major "political parties" that explicitly bribe coin holders to vote for them, and also require each member in the pool to vote for all the others.

Here's LiskElite, with fifty-five members (out of a total 101):

:cure | https://liskelite.com

Lisk

Member Rules:

1. Every member of Elite except the china delegate must share 25% of his/her forging LISK to his/her Voters every week;

2. Every member of Elite except the china delegate must donate 5% of forging LISK to the Elite Lisk fund used to support Lisk ecosystem;

3. Every member of Elite must vote for other members;

4. Elite membership registration is now closed and no new members are currently accepted.

Voter Rules:

1. For getting the rewards you must vote for all of Elite Group members;

2. Elite reward payouts will be done on a weekly basis and will be paid out to voter accounts automatically.

All rights reserved by Elite Group

Here's LiskGDT, with thirty-three members:

pool.liskgdt.net

And as "Exhibit B" some voter bribes being paid out in Ark:

Latest Transactions

Id	Timestamp	Sender	Recipient	Smartbridge	Amount (ARK)	Fee (ARK)
380af...d7ab4	2017/04/17 12:20:41	bioly	AbxqF...jXJhB	Payout from bioly delegate pool, thank you for support!	7.60466706	0.1
5795e...26029	2017/04/17 12:20:41	bioly	ARUNS...oLzvs	Payout from bioly delegate pool, thank you for support!	6.07691376	0.1
37694...35419	2017/04/17 12:20:40	bioly	AG2Ni1...taeiZv	Payout from bioly delegate pool, thank you for support!	2.48455539	0.1
8c6b1...f1f9a	2017/04/17 12:20:39	bioly	AWmMj...HJU8R	Payout from bioly delegate pool, thank you for support!	118.47841646	0.1
d2ad5...c84af	2017/04/17 12:20:38	bioly	AbJ6N...ZrZxq	Payout from bioly delegate pool, thank you for support!	9.37653981	0.1
45280...aa3f0	2017/04/17 12:20:37	bioly	AevZb...68d6C	Payout from bioly delegate pool, thank you for support!	118.4945548	0.1
ace28...1cdee	2017/04/17 12:20:37	bioly	teletobi	Payout from bioly delegate pool, thank you for support!	11.72867675	0.1
20ca3...4278b	2017/04/17 12:20:36	bioly	ANY7W...6TfzX	Payout from bioly delegate pool, thank you for support!	4.80016674	0.1
a4de1...f90fd	2017/04/17 12:20:36	bioly	ARKlla...znvzZ	Payout from bioly delegate pool, thank you for support!	178.80073745	0.1
cb528...592bc	2017/04/17 12:20:36	bioly	ALtmaL...QeHyP	Payout from bioly delegate pool, thank you for support!	237.32335576	0.1
29740...578db	2017/04/17 12:20:35	bioly	AUw4A...HxWB7	Payout from bioly delegate pool, thank you for support!	54.14948207	0.1
331df...5b0f2	2017/04/17 12:20:35	bioly	AQxnW...F2HGH	Payout from bioly delegate pool, thank you for support!	46.96456749	0.1
38fac...e02f5	2017/04/17 12:20:34	bioly	AXKvf...L8TW9	Payout from bioly delegate pool, thank you for support!	41.98709123	0.1
50190...b52B4	2017/04/17 12:20:34	bioly	AWukK...bR84m	Payout from bioly delegate pool, thank you for support!	7.39663982	0.1
72270...78s41	2017/04/17 12:20:34	bioly	AUTPB...E6pro	Payout from bioly delegate pool, thank you for support!	15.64031609	0.1
199f4...bae6a	2017/04/17 12:20:33	bioly	AVbiK...MEK8P	bioly fee account	403.66128558	0.1
af13f...6a16e	2017/04/17 12:20:33	bioly	AVVVY...gTtCo	Payout from bioly delegate pool, thank you for support!	7.63884129	0.1
74c3c...061f6	2017/04/17 12:20:32	bioly	AYTAy...3egy6	Payout from bioly delegate pool, thank you for support!	71.46381847	0.1

Here, note that there is a key difference between tightly coupled and loosely coupled votes. In a loosely coupled vote, direct or indirect vote bribing is also possible, but if the community agrees that some given proposal or set of votes constitutes a game-theoretic attack, they can simply socially agree to ignore it. And in fact this has kind of already happened—the Carbonvote contains a blacklist of addresses corresponding to known exchange addresses, and votes from these addresses are not counted. In a tightly coupled vote, there is no way to create such a blacklist at protocol level, because agreeing who is part of the blacklist is *itself* a blockchain-governance decision. But since the blacklist is part of a community-created voting tool that only indirectly influences protocol changes, voting tools that contain bad blacklists can simply be rejected by the community.

It's worth noting that this section **is not** a prediction that all tightly coupled voting systems will quickly succumb to bribe attacks. It's entirely possible that many will survive for one simple reason: all of these projects have founders or foundations with large premines, and these act as large centralized actors that are interested in their platforms' success that are not vulnerable to bribes, and hold enough coins to outweigh most bribe attacks. However, this kind of centralized-trust model, while arguably useful in some contexts in a project's early stages, is clearly one that is not sustainable in the long term.

NON-REPRESENTATIVENESS

Another important objection to voting is that coin holders are only one class of user, and may have interests that collide with those of other users. In the case of pure cryptocurrencies like

Bitcoin, store-of-value use ("hodling")* and medium-of-exchange use ("buying coffees") are naturally in conflict, as the store-of-value use case prizes security much more than the medium-of-exchange use case, which more strongly values usability. With Ethereum, the conflict is worse, as there are many people who use Ethereum for reasons that have nothing to do with ether (see: CryptoKitties), or even value-bearing digital assets in general (see: ENS).

Additionally, even if coin holders are the only relevant class of user (one might imagine this to be the case in a cryptocurrency where there is an established social contract whose purpose is to be the next digital gold, and nothing else), there is still the challenge that a coin-holder vote gives a much greater voice to wealthy coin holders than to everyone else, opening the door for centralization of holdings to lead to unencumbered centralization of decision-making. Or in other words . . .

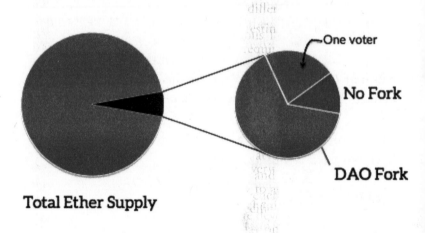

* "HODL" is a term of art in the crypto lexicon that refers to someone furiously trying to type "hold" to keep others from selling when a token's price drops. It is a rallying cry most associated with price-focused traders; in the culture of Ethereum, the corresponding meme is "BUIDL," a call to respond to setbacks by building better, more usable tools.

This criticism applies to both tightly coupled and loosely coupled voting equally; however, loosely coupled voting is more amenable to compromises that mitigate its unrepresentativeness, and we will discuss this more later.

CENTRALIZATION

Let's look at the existing live experiment that we have in tightly coupled voting on Ethereum, the gas limit. Here's the gas limit evolution over the past two years:

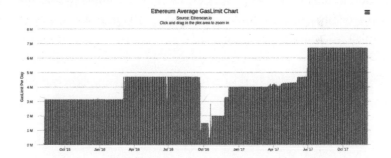

You might notice that the general feel of the curve is a bit like another chart that may be quite familiar to you:

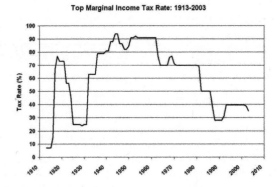

Basically, they both look like magic numbers that are created and repeatedly renegotiated by a fairly centralized group of guys sitting together in a room. What's happening in the first case? Miners are generally following the direction favored by the community, which is itself gauged via social consensus aids similar to those that drive hard forks (core-developer support, Reddit upvotes, etc.; in Ethereum, the gas limit has never gotten controversial enough to require anything as serious as a coin vote).

Hence, it is not at all clear that voting will be able to deliver results that are actually decentralized, if voters are not technically knowledgeable and simply defer to a single dominant tribe of experts. This criticism once again applies to tightly coupled and loosely coupled voting equally.

UPDATE: Since writing this, it seems like Ethereum miners managed to up the gas limit from 6.7 million to 8 million all without even discussing it with the core developers or the Ethereum Foundation. So there is hope; but it takes a lot of hard community building and other grueling non-technical work to get to that point.

DIGITAL CONSTITUTIONS

One approach that has been suggested to mitigate the risk of runaway bad governance algorithms is "digital constitutions" that mathematically specify desired properties that the protocol should have, and require any new code changes to come with a computer-verifiable proof that they satisfy these properties. This seems like a good idea at first, but this, too, should, in my opinion, be viewed skeptically.

In general, the idea of having norms about protocol properties, and having these norms serve the function of one of the coordination flags, is a very good one. This allows us to enshrine core properties of a protocol that we consider to be very important and valuable, and make them more difficult to change. However, this is exactly the sort of thing that should be enforced in loosely coupled (layer 2), rather than tightly coupled (layer 1), form.

Basically any meaningful norm is actually quite hard to express in its entirety; this is part of the complexity of the value problem. This is true even for something as seemingly unambiguous as the 21-million-coin limit.* Sure, one can add a line of code saying `assert total_supply <= 21000000`, and put a comment around it saying "do not remove at all costs," but there are plenty of roundabout ways of doing the same thing. For example, one could imagine a soft fork that adds a mandatory transaction fee that is proportional to coin value × time since the coins were last sent, and this is equivalent to demurrage, which is equivalent to deflation. One could also implement another currency, called Bjtcoin, with 21 million *new* units, and add a feature where if a bitcoin transaction is sent the miner can intercept it and claim the bitcoin, instead giving the recipient bjtcoin; this would rapidly force bitcoins and bjtcoins to be fungible with each other, increasing the "total supply" to 42 million without ever tripping up that line of code. "Softer" norms like not interfering with application state are even harder to enforce.

We *want* to be able to say that a protocol change that violates any of these guarantees should be viewed as illegitimate—there should be a coordination institution that waves a red flag—even if they get approved by a vote. We also want to be able to say a protocol change that follows the letter of a norm, but blatantly violates its spirit, should *still* be viewed as illegitimate. And having norms exist on layer 2—in the minds of humans in

---

* A reference to the number of total coins the Bitcoin system will produce under its current design.

the community, rather than in the code of the protocol—best achieves that goal.

TOWARD A BALANCE

However, I am also not willing to go the other way and say that coin voting, or other explicit on-chain-voting-like schemes, have no place in governance whatsoever. The leading alternative seems to be core-developer consensus, however the failure mode of a system being controlled by "ivory tower intellectuals" who care more about abstract philosophies and solutions that sound technically impressive over and above real day-to-day concerns like user experience and transaction fees is, in my view, also a real threat to be taken seriously.

So how do we solve this conundrum? Well, first, we can heed the words of slatestarcodex* in the context of traditional politics:

> The rookie mistake is: you see that some system is partly Moloch [i.e., captured by misaligned special interests], so you say "Okay, we'll fix that by putting it under the control of this other system. And we'll control this other system by writing 'DO NOT BECOME MOLOCH' on it in bright red marker."
>
> ("I see capitalism sometimes gets misaligned. Let's fix it by putting it under control of the government. We'll control the government by having only virtuous people in high offices.") I'm not going to claim there's a great

---

* *Slate Star Codex* is the blog of Scott Alexander, "a psychiatrist on the US West Coast." His blog is widely read in crypto culture. The essay "Meditations on Moloch" interprets the ancient Levantine child-eating god, through the lens of Allen Ginsberg's poem "Howl," as a god of coordination failure. In the Ethereum subculture, "slaying Moloch" is a byword for building a better system for coordination through aligned incentives.

alternative, but the occasionally adequate alternative is the neoliberal one—find a couple of elegant systems that all optimize along different criteria approximately aligned with human happiness, pit them off against each other in a structure of checks and balances, hope they screw up in different places like in that swiss cheese model, keep enough individual free choice around that people can exit any system that gets too terrible, and let cultural evolution do the rest.

In blockchain governance, it seems like this is the only way forward as well. The approach for blockchain governance that I advocate is "multifactorial consensus," where different coordination flags and different mechanisms and groups are polled, and the ultimate decision depends on the collective result of all these mechanisms together. These coordination flags may include:

□   The road map (i.e., the set of ideas broadcasted earlier on in the project's history about the direction in which the project would be going)

□   Consensus among the dominant core-development teams

□   Coin-holder votes

□   User votes, through some kind of Sybil-resistant polling system

□   Established norms (e.g., non-interference with applications, the 21-million-coin limit)

I would argue that it is very useful for coin voting to be one of several coordination institutions deciding whether or not a given change gets implemented. It is an imperfect and unrepresentative signal, but it is a *Sybil-resistant** one—if you see 10 million ETH voting for a given proposal, you *cannot* dismiss that by simply saying, "Oh, that's just hired Russian trolls with fake social media accounts." It is also a signal that is sufficiently distinct from the core development team that, if needed, it can serve as a check on it. However, as described above, there are very good reasons why it should not be the *only* coordination institution.

And underpinning it all is the key difference from traditional systems that makes blockchains interesting: the "layer 1" that underpins the whole system is the requirement for individual users to assent to any protocol changes, and their freedom, and credible threat, to "fork off" if someone attempts to force changes on them that they consider hostile.

Tightly coupled voting is also okay to have in some limited contexts—for example, despite its flaws, miners' ability to vote on the gas limit is a feature that has proven very beneficial on multiple occasions. The risk that miners will try to abuse their power may well be lower than the risk that any specific gas limit or block-size limit hard-coded by the protocol on day one will end up leading to serious problems, and in that case letting miners vote on the gas limit is a good thing. However, "allowing miners or validators to vote on a few specific parameters that need to be rapidly changed from time to time" is a very far cry from giving them arbitrary control over protocol rules, or letting voting control validation, and these more expansive visions of on-chain governance have a much murkier potential, both in theory and in practice.

---

* Sybil resistance is the property of addressing a potential "Sybil attack," when a single user is able to undermine a system by posing as multiple users. The name is a reference to the best-selling 1973 book *Sybil*, which purported to be an account of what was then called "multiple personality disorder."

ON COLLUSION

vitalik.ca
April 3, 2019

Over the last few years there has been an increasing interest in using deliberately engineered economic incentives and mechanism design to align behavior of participants in various contexts. In the blockchain space, mechanism design first and foremost provides the security for the blockchain itself, encouraging miners or proof-of-stake validators to participate honestly, but more recently it is being applied in prediction markets, "token curated registries," and many other contexts. The nascent RadicalxChange movement has meanwhile spawned experimentation with Harberger taxes, quadratic voting, quadratic financing, and more. More recently, there has also been growing interest in using token-based incentives to try to encourage quality posts in social media. However, as development of these systems moves from theory to practice, there are a number of challenges that need to be addressed, challenges that I would argue have not yet been adequately confronted.

A recent example of this move from theory toward deployment is Bihu, a Chinese platform that has recently released a coin-based

143

mechanism for encouraging people to write posts. The basic mechanism is that if a user of the platform holds KEY tokens, they have the ability to stake those KEY tokens on articles; every user can make k "upvotes" per day, and the "weight" of each upvote is proportional to the stake of the user making the upvote. Articles with a greater quantity of stake upvoting them appear more prominently, and the author of an article gets a reward of KEY tokens roughly proportional to the quantity of KEY upvoting that article. This is an oversimplification and the actual mechanism has some nonlinearities baked into it, but they are not essential to the basic functioning of the mechanism. KEY has value because it can be used in various ways inside the platform, but particularly a percentage of all ad revenues gets used to buy and burn KEY (yay, big thumbs up to them for doing this and not making yet another medium-of-exchange token!).

This kind of design is far from unique; incentivizing online content creation is something that very many people care about, and there have been many designs of a similar character, as well some fairly different designs. And in this case this particular platform is already being used significantly:

A few months ago, the Ethereum-trading subreddit /r/eth-trader introduced a somewhat similar experimental feature where a token called "donuts" is issued to users that make comments that get upvoted, with a set amount of donuts issued weekly to users in proportion to how many upvotes their comments received. The donuts could be used to buy the right to set the contents of the banner at the top of the subreddit, and could also be used to vote in community polls. However, unlike what happens in the KEY system, here the reward that B receives when B is upvoted by A is not proportional to A's existing coin supply; instead, each Reddit account has an equal ability to contribute to other Reddit accounts.

These kinds of experiments, attempting to reward quality content creation in a way that goes beyond the known limitations of donations and microtipping, are very valuable; under-compensation of user-generated internet content is a very significant problem in society in general, and it's heartening to see crypto communities attempting to use the power of mechanism design to make inroads on solving it. **But unfortunately, these systems are also vulnerable to attack.**

SELF-VOTING, PLUTOCRACY, AND BRIBES

Here is how one might economically attack the design proposed above. Suppose that some wealthy user acquires some quantity n of tokens, and as a result each of the user's k upvotes gives the recipient a reward of $n \times q$ (q here probably being a very small number—e.g., think $q = 0.000001$). The user simply upvotes their own sockpuppet* accounts, giving themselves the reward of $n \times k \times q$. Then, the system simply collapses into each user having an "interest rate" of $k \times q$ per period, and the mechanism accomplishes nothing else.

The actual Bihu mechanism seemed to anticipate this, and has some superlinear logic where articles with more KEY upvoting them gain a disproportionately greater reward, seemingly to encourage upvoting popular posts rather than self-upvoting. It's a common pattern among coin-voting governance systems to add this kind of superlinearity to prevent self-voting from undermining the entire system; most DPoS schemes have a limited number of delegate slots with zero rewards for anyone who does not get enough votes to join one of the slots, with similar effect. But these schemes invariably introduce two new weaknesses:

☐ They **subsidize plutocracy**, as very wealthy individuals and cartels can still get enough funds to self-upvote.

☐ They can be circumvented by users **bribing** other users to vote for them en masse.

Bribing attacks may sound farfetched (who here has ever accepted

---

* A sockpuppet account is a fake account a user creates while purporting it to be a different user.

a bribe in real life?), but in a mature ecosystem they are much more realistic than they seem. In most contexts where bribing has taken place in the blockchain space, the operators use a euphemistic new name to give the concept a friendly face: it's not a bribe, it's a "staking pool" that "shares dividends." Bribes can even be obfuscated: imagine a cryptocurrency exchange that offers zero fees and spends the effort to make an abnormally good user interface, and does not even try to collect a profit; instead, it uses coins that users deposit to participate in various coin-voting systems. There will also inevitably be people that see in-group collusion as just plain normal; see a recent scandal involving EOS DPoS for one example:

Finally, there is the possibility of a "negative bribe" (i.e., blackmail or coercion), threatening participants with harm unless they act inside the mechanism in a certain way.

In the /r/ethtrader experiment, fear of people coming in and *buying* donuts to shift governance polls led to the community deciding to make only locked (i.e., untradeable) donuts eligible for use in voting. But there's an even cheaper attack than buying donuts (an attack that can be thought of as a kind of obfuscated bribe): *renting* them. If an attacker is already holding ETH, they can use it as collateral on a platform like Compound to take out a loan of some token, giving you the full right to use that token for whatever purpose including participating in votes, and when they're done they simply send the tokens back to the loan contract to get their collateral back—all without having to endure even a second of price exposure to the token that they just used to swing a coin vote, even if the coin-vote mechanism includes a time lockup (as Bihu does). In every case, issues around bribing, and accidentally over-empowering well-connected and wealthy participants, prove surprisingly difficult to avoid.

IDENTITY

Some systems attempt to mitigate the plutocratic aspects of coin voting by making use of an identity system. In the case of the /r/ethtrader donut system, for example, although *governance polls* are done via coin vote, the mechanism that determines *how many donuts* (i.e., coins) you get in the first place is based on Reddit accounts: 1 upvote from one Reddit account = *n* donuts earned. The ideal goal of an identity system is to make it relatively easy for individuals to get one identity, but relatively difficult to get many identities. In the /r/ethtrader donut system, that's Reddit accounts, in the Gitcoin CLR matching gadget* it's GitHub accounts that are used for the same purpose. But identity, at least the way it has been implemented so far, is a fragile thing . . .

---

* Gitcoin is a funding platform for building open-source software, particularly in the Ethereum ecosystem. The CLR mechanism was an experiment in distributing matching funds with community donations, following the concept of "quadratic funding" proposed by Buterin, Zoë Hitzig, and E. Glen Weyl.

Jamie Bartlett @
@JamieJBartlett
Follow ∨

I'm completely obsessed by click farms -
where thousands of machines are lined
up to generate fake engagement.

🌐 English Russia

0:32 · 4M views

10:00 AM - 11 Mar 2019

21,561 Retweets 42,339 Likes

💬 1.5K 🔁 22K ❤ 42K ✉

Oh, are you too lazy to make a big rack of phones? Well maybe you're looking for this:

BuyAccs.com
СЕРВИС РЕГИСТРАЦИИ АККАУНТОВ

Russian Version English Version

Наш магазин аккаунтов рад предложить аккаунты различных **почтовых**
получаете аккаунты **СРАЗУ** *после оплаты* заказа. Мы принимаем **крип'**
еще около 30 платежных систем через **Unitpay.ru.**

При покупке аккаунтов менее 1000 штук действует специальный тариф.

Заработай на продаже аккаунтов

Купить аккаунты Одноклассников
Купить аккаунты Вконтакте
Купить аккаунты Мамба

💰 **Сейчас в продаже**

Служба	В наличии	Цена за 1К аккаунтов
Mail.ru	475698	1K-10K: **$7** | 10K-20K: **$6.5** | 20K+: **$6**
Yandex.ru	16775	1K-10K: **$50** | 10K-20K: **$50** | 20K+: **$50**
Rambler.ru	6694	1K-10K: **$30** | 10K-20K: **$30** | 20K+: **$30**
Rambler.ru Mix	8037	1K-10K: **$30** | 10K-20K: **$30** | 20K+: **$30**
Rambler.ru Promo	176605	1K-10K: **$6** | 10K-20K: **$5.5** | 20K+: **$5**
Bigmir.net	10000	1K-10K: **$18** | 10K-20K: **$18** | 20K+: **$18**
I.ua	14020	1K-10K: **$18** | 10K-20K: **$17** | 20K+: **$16**
Gmail.com 2015 USA	2326	1K-10K: **$450** | 10K-20K: **$450** | 20K+: **$450**
Gmail.com 2015 USA PVA	6504	1K-10K: **$800** | 10K-20K: **$800** | 20K+: **$800**

Usual warning about how sketchy sites may or may not scam you, do your own research, etc., etc. applies.

Arguably, attacking these mechanisms by simply controlling thousands of fake identities like a puppet master is *even easier* than having to go through the trouble of bribing people. And if you think the response is to just increase security to go up to *government-level* IDs? Keep in mind that there are specialized criminal organizations that are well ahead of you, and even if all the underground ones are taken down, hostile governments are definitely going to create fake passports by the millions if we're stupid enough to create systems that make that sort of activity profitable. And this doesn't even begin to mention attacks in the opposite direction, identity-issuing institutions attempting to disempower marginalized communities by *denying* them identity documents . . .

COLLUSION

Given that so many mechanisms seem to fail in such similar ways once multiple identities or even liquid markets get into the picture, one might ask, is there some deep common strand that causes all of these issues? I would argue the answer is yes, and the "common strand" is this: it is much harder, and more likely to be outright impossible, to make mechanisms that maintain desirable properties in a model where participants can collude, than in a model where they can't. Most people likely already have some intuition about this; specific instances of this principle are behind well-established norms and often laws promoting competitive markets and restricting price-fixing cartels, vote buying and selling, and bribery. But the issue is much deeper and more general.

In the version of game theory that focuses on individual choice—that is, the version that assumes that each participant makes decisions independently and that does not allow for the possibility of groups of agents working as one for their mutual benefit—there are mathematical proofs that at least one stable Nash equilibrium must exist in any game, and mechanism designers have a very wide latitude to "engineer" games to achieve specific outcomes. But in the version of game theory that allows for the possibility of coalitions working together, called *cooperative game theory*, **there are large classes of games that do not have any stable outcome that a coalition cannot profitably deviate from**.

Majority games, formally described as games of n agents where any subset of more than half of them can capture a fixed reward and split it among themselves, a setup eerily similar to many situations in corporate governance, politics, and other situations in human life, are part of that set of inherently unstable games. That is to say, if there is a situation with some fixed pool of resources and some currently established mechanism for distributing those resources, and it's unavoidably possible for 51% of the participants to conspire to seize control of the resources, no matter what the current configuration is, there is always some conspiracy that can emerge that would be profitable for the participants. However, that conspiracy would then in turn be vulnerable to potential new conspiracies, possibly including a combination of previous conspirators and victims . . . and so on and so forth.

Round	A	B	C
1	1/3	1/3	1/3
2	1/2	1/2	0
3	2/3	0	1/3
4	0	1/3	2/3

This fact, the instability of majority games under cooperative game theory, is arguably highly underrated as a simplified general mathematical model of why there may well be no "end of history" in politics and no system that proves fully satisfactory; I personally believe it's much more useful than the more famous Arrow's theorem,* for example.

There are two ways to get around this issue. The first is to try to restrict ourselves to the class of games that *are* "identity-free" and "collusion-safe," so where we do not need to worry about either bribes or identities. The second is to try to attack the identity and collusion-resistance problems directly, and actually solve them well enough that we can implement non-collusion-safe games with the richer properties that they offer.

IDENTITY-FREE AND COLLUSION-SAFE GAME DESIGN

The class of games that is identity-free and collusion-safe is substantial. Even proof of work is collusion-safe up to the bound of a single actor having about 23.21% of total hashpower, and this bound can be increased up to 50% with clever engineering. Competitive markets are reasonably collusion-safe up until a relatively high bound, which is easily reached in some cases but in other cases is not.

In the case of *governance* and *content curation* (both of which are really just special cases of the general problem of identifying public goods and public bads), a major class of mechanism that works well is *futarchy*—typically portrayed as "governance by prediction market," though I would also argue that the use of security deposits is fundamentally in the same class of technique. The way futarchy mechanisms, in their most general form, work is that they make "voting" not just an expression of opinion, but also a *prediction*, with a reward for making predictions that are true and a penalty for making predictions that are false. For example, my proposal for "prediction markets for content curation DAOs" suggests

---

* A mathematical finding published by Kenneth Arrow in 1951 about the impossibility of achieving a set of desirable results through ranked-choice voting systems.

a semi-centralized design where anyone can upvote or downvote submitted content, with content that is upvoted more being more visible, where there is also a "moderation panel" that makes final decisions. For each post, there is a small probability (proportional to the total volume of upvotes and downvotes on that post) that the moderation panel will be called on to make a final decision on the post. If the moderation panel approves a post, everyone who upvoted it is rewarded and everyone who downvoted it is penalized, and if the moderation panel disapproves a post the reverse happens; this mechanism encourages participants to make upvotes and downvotes that try to "predict" the moderation panel's judgments.

Another possible example of futarchy is a governance system for a project with a token, where anyone who votes for a decision is obligated to purchase some quantity of tokens at the price at the time the vote begins if the vote wins; this ensures that voting on a bad decision is costly, and in the limit if a bad decision wins a vote everyone who approved the decision must essentially buy out everyone else in the project. This ensures that an individual vote for a "wrong" decision can be very costly for the voter, precluding the possibility of cheap bribe attacks.

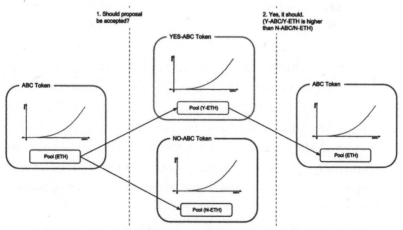

A graphical description of one form of futarchy, creating two markets representing the two "possible future worlds" and picking the one with a more favorable price.

However, the range of things that mechanisms of this type can do is limited. In the case of the content curation example above, we're not really solving governance, we're just scaling the functionality of a governance gadget that is already assumed to be trusted. One could try to replace the moderation panel with a prediction market on the price of a token representing the right to purchase advertising space, but in practice prices are too noisy an indicator to make this viable for anything but a very small number of very large decisions. And often the value that we're trying to maximize is explicitly something other than the maximum value of a coin.

Let's take a more explicit look at why, in the more general case where we can't easily determine the value of a governance decision via its impact on the price of a token, good mechanisms for identifying public goods and bads unfortunately cannot be identity-free or collusion-safe. If one tries to preserve the property of a game being identity-free, building a system where identities don't matter and only coins do, there is an impossible tradeoff between either failing to incentivize legitimate public goods or over-subsidizing plutocracy.

The argument is as follows. Suppose that there is some author that is producing a public good (e.g., a series of blog posts) that provides value to each member of a community of ten thousand people. Suppose there exists some mechanism where members of the community can take an action that causes the author to receive a gain of one dollar. Unless the community members are *extremely* altruistic, for the mechanism to work the cost of taking this action must be much lower than one dollar, otherwise the portion of the benefit captured by the member of the community supporting the author would be much smaller than the cost of supporting the author, and so the system collapses into a tragedy of the commons where no one supports the author. Hence, there must exist a way to cause the author to earn one dollar at a cost much less than one dollar. But now suppose that there is also a fake community, which consists of ten thousand fake sockpuppet accounts of the same wealthy attacker. This community takes all of the same actions as the real community, except instead of supporting the author, they support *another fake* account, which is also a sockpuppet of the attacker. If it was possible for a member of the "real community" to give the author one dollar at a personal cost of much less than one dollar, it's possible for the attacker to give *themselves* one dollar at a cost much less than one dollar over and over again, and thereby drain the system's funding. Any mechanism that can help genuinely under-coordinated parties coordinate will, without the right safeguards, also help already coordinated parties (such as many accounts controlled by the same person) *over-coordinate*, extracting money from the system.

A similar challenge arises when the goal is not funding but determining what content should be most visible. What content do you think would get more dollar value supporting it: A legitimately high-quality blog article benefiting thousands of people but benefiting each individual person relatively slightly, or this?

Or perhaps this?*

Those who have been following recent politics "in the real world" might also point out a different kind of content that benefits highly centralized actors: social media manipulation by hostile governments. Ultimately, both centralized systems and decentralized systems are facing the same fundamental problem, which is that **the "marketplace of ideas" (and of public goods more generally) is very far from an "efficient market" in the sense that economists normally use the term**, and this leads to both underproduction of public goods even in "peacetime" but also vulnerability to active attacks. It's just a hard problem.

This is also why coin-based voting systems (like Bihu's) have one major genuine advantage over identity-based systems (like the Gitcoin CLR or the /r/ethtrader donut experiment): at least there is no benefit to buying accounts en masse, because everything you do is proportional to how many coins you have, regardless of how many accounts the coins are split between. However, mechanisms that do not rely on any model of identity and rely only on coins fundamentally cannot solve the problem of concentrated interests outcompeting dispersed communities trying to support public

---

* Bitconnect was a cryptocurrency investment platform that shut down in 2018 after regulators began scrutinizing it for being a Ponzi scheme.

goods; an identity-free mechanism that empowers distributed communities cannot avoid over-empowering centralized plutocrats pretending to be distributed communities.

But it's not just identity issues that public goods games are vulnerable to; it's also bribes. To see why, consider again the example above, but where instead of the "fake community" being 10,001 sockpuppets of the attacker, the attacker only has one identity, the account receiving funding, and the other ten thousand accounts are real users—but users that receive a bribe of one cent each to take the action that would cause the attacker to gain an additional one dollar. As mentioned above, these bribes can be highly obfuscated, even through third-party custodial services that vote on a user's behalf in exchange for convenience, and in the case of "coin vote" designs an obfuscated bribe is even easier: one can do it by renting coins on the market and using them to participate in votes. Hence, while some kinds of games, particularly prediction market or security-deposit-based games, can be made collusion-safe and identity-free, generalized public-goods funding seems to be a class of problem where collusion-safe and identity-free approaches unfortunately just cannot be made to work.

COLLUSION RESISTANCE AND IDENTITY

The other alternative is attacking the identity problem head on. As mentioned above, simply going up to higher-security centralized identity systems, like passports and other government IDs, will not work at scale; in a sufficiently incentivized context, they are very insecure and vulnerable to the issuing governments themselves! Rather, the kind of "identity" we are talking about here is some kind of robust multifactorial set of claims that an actor identified by some set of messages actually is a unique individual.

A very early proto-model of this kind of networked identity is arguably social recovery in HTC's blockchain phone:

The basic idea is that your private key is secret-shared between up to five trusted contacts, in such a way that mathematically ensures that three of them can recover the original key, but two or fewer can't. This qualifies as an "identity system"—it's your five friends determining whether or not someone trying to recover your account actually is you. However, it's a special-purpose identity system trying to solve a problem—personal account security—that is different from (and easier than!) the problem of attempting to identify unique humans. That said, the general model of individuals making claims about each other can quite possibly be bootstrapped into some kind of more robust identity model. These systems could be augmented if desired using the "futarchy" mechanic described above: if someone makes a claim that someone is a unique human, and someone else disagrees, and both sides are willing to put down a bond to litigate the issue, the system can call together a judgment panel to determine who is right.

But we also want another crucially important property: we want an identity that you cannot credibly rent or sell. Obviously, we can't prevent people from making a deal ("you send me fifty, I'll send you my key"), but what we *can* try to do is prevent such deals from being *credible*—make it so that the seller can easily cheat the buyer and give the buyer a key that doesn't actually work. One way to do this is to make a mechanism by which the owner of a key can send a transaction that revokes the key and replaces it with another key of the owner's choice, all in a way that cannot be proven. Perhaps the simplest way to get around this is to either use a trusted party that runs the computation and only publishes results (along with zero-knowledge proofs proving the results, so the trusted party is trusted only for privacy, not integrity), or decentralize the same functionality through multiparty computation. Such approaches will not solve collusion completely; a group of friends could still come together and sit on the same couch and coordinate votes. But collusion can be reduced to a manageable extent that will not lead to these systems outright failing.

There is a further problem: initial distribution of the key. What happens if a user creates their identity inside a third-party custodial service that then stores the private key and uses it to clandestinely make votes on things? This would be an implicit bribe, the user's voting power in exchange for providing to the user a convenient service, and what's more, if the system is secure in that it successfully prevents bribes by making votes unprovable, clandestine voting by third-party hosts would *also* be undetectable. The only approach that gets around this problem seems to be . . . in-person verification. For example, one could have an ecosystem of "issuers" where each issuer issues smart cards with private keys, which the user can immediately download onto their smartphone and send a message to replace the key with a different key that they do not reveal to anyone. These issuers could be meetups and conferences, or potentially individuals that have already been deemed by some voting mechanic to be trustworthy.

Building out the infrastructure for making collusion-resistant mechanisms possible, including robust decentralized identity systems, is a difficult challenge, but if we want to unlock the potential of such mechanisms, it seems unavoidable that we have to do our best to try. It is true that the current computer-security dogma around, for example, introducing online voting is simply "don't," but if we want to expand the role of voting-like mechanisms, including more advanced forms such as quadratic voting and quadratic finance, to more roles, we have no choice but to confront the challenge head on, try really hard, and hopefully succeed at making something secure enough for at least some use cases.

ON FREE SPEECH

---

vitalik.ca
April 16, 2019

A statement may be both true and dangerous.
The previous sentence is such a statement.
—DAVID FRIEDMAN

Freedom of speech is a topic that many internet communities have struggled with over the last two decades. Cryptocurrency and blockchain communities, a major part of their raison d'être being censorship resistance, are especially poised to value free speech very highly, and yet, over the last few years, the extremely rapid growth of these communities and the very high financial and social stakes involved have repeatedly tested the application and the limits of the concept. In this post, I aim to disentangle some of the contradictions, and make a case for what the norm of "free speech" really stands for.

"FREE SPEECH LAWS" VS. "FREE SPEECH"

A common, and in my own view frustrating, argument that I often hear is that "freedom of speech" is exclusively a legal restriction on what governments can act against, and has nothing to say regarding the actions of private entities such as corporations, privately owned platforms, internet forums, and conferences. One of the larger examples of "private censorship" in cryptocurrency communities was the decision of Theymos, the moderator of the /r/bitcoin subreddit, to start heavily moderating the subreddit, forbidding arguments in favor of increasing the Bitcoin blockchain's transaction capacity via a hard fork.

[–] **theymos** -45 points 1 year ago*

You can promote BIP 101 as an idea. You can't promote (on /r/Bitcoin) the actual *usage* of BIP 101. When the idea has consensus, *then* it can be rolled out.

Bitcoin is not a democracy. Not of miners, and not of nodes. Switching to XT is not a vote for BIP 101 -- it is abandoning Bitcoin for a separate network/currency. It is good that you have the freedom to do this. One of the great things about Bitcoin *is* its lack of democracy: even if 99% of people use Bitcoin, you are free to implement BIP 101 in a separate currency without the Bitcoin users being able to democratically coerce you into using the real Bitcoin network/currency again. But I am not obligated to allow these separate offshoots of Bitcoin to exist on /r/Bitcoin, and I'm not going to.

A common strategy used by defenders of Theymos's censorship was to say that heavy-handed moderation is okay because /r/bitcoin is "a private forum" owned by Theymos, and so he has the right to do whatever he wants in it; those who dislike it should move to other forums:

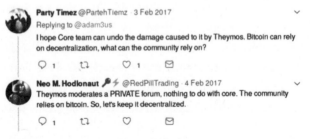

Party Timez @PartehTiemz · 3 Feb 2017
Replying to @adam3us
I hope Core team can undo the damage caused to it by Theymos. Bitcoin can rely on decentralization, what can the community rely on?
1 · 1

Neo M. Hodlonaut @RedPillTrading · 4 Feb 2017
Theymos moderates a PRIVATE forum, nothing to do with core. The community relies on bitcoin. So, let's keep it decentralized.
1

▲ beaner 6 months ago [-]

Bitcoin cash isn't censored. It has its own subreddit (and the rest of the internet) where discussion can be had about it.

Equating "censored in r/bitcoin" with censorship in general sort of proves that it's mostly about politics; you want to be uncensored _in a specific private community_. If BCH can stand on its own merit (and hopefully it can!) then you don't need that. Those who think it does need that aren't trying to make BCH successful, they want to control Bitcoin. And so it makes sense that people with those motives should not be allowed.

Layer 2 is a scaling solution, I don't see why it wouldn't be.

And it's true that Theymos has not *broken any laws* by moderating his forum in this way. But to most people, it's clear that there is still some kind of free speech violation going on. So what gives? First of all, it's crucially important to recognize that freedom of speech is not just a *law in some countries*. It's also a social principle. And the underlying goal of the social principle is the same as the underlying goal of the law: to foster an environment where the ideas that win are ideas that are good, rather than just the ideas that happen to be favored by people in a position of power. And governmental power is not the only kind of power that we need protection from; there is also a corporation's power to fire someone, an internet forum moderator's power to delete almost every post in a discussion thread, and many other kinds of power both hard and soft.

So what is the underlying social principle here? Quoting Eliezer Yudkowsky:*

> There are a very few injunctions in the human art of rationality that have no ifs, ands, buts, or escape clauses. This is one of them. Bad argument gets counterargument. Does not get bullet. Never. Never ever never for ever.

Slatestarcodex elaborates:**

> What does "bullet" mean in the quote above? Are other projectiles covered? Arrows? Boulders launched from catapults? What about melee weapons like swords or maces? Where exactly do we draw the line for "inappropriate responses to

---

* See note above in "On Silos."
** See note above in "Notes on Blockchain Governance."

an argument"? A good response to an argument is one that addresses an idea; a bad argument is one that silences it. If you try to address an idea, your success depends on how good the idea is; if you try to silence it, your success depends on how powerful you are and how many pitchforks and torches you can provide on short notice. Shooting bullets is a good way to silence an idea without addressing it. So is firing stones from catapults, or slicing people open with swords, or gathering a pitchfork-wielding mob. But trying to get someone fired for holding an idea is also a way of silencing an idea without addressing it.

That said, sometimes there is a rationale for "safe spaces" where people who, for whatever reason, just don't want to deal with arguments of a particular type can congregate and where those arguments actually do get silenced. Perhaps the most innocuous of all are spaces like ethresear.ch where posts get silenced just for being "off topic" to keep the discussion focused. But there's also a dark side to the concept of "safe spaces"; as Ken White writes:*

> This may come as a surprise, but I'm a supporter of "safe spaces." I support safe spaces because I support freedom of association. Safe spaces, if designed in a principled way, are just an application of that freedom . . . But not everyone imagines "safe spaces" like that. Some use the concept of "safe spaces" as a sword, wielded to annex public spaces and demand that people within those spaces conform to their private norms. That's not freedom of association.

---

* Ken White is a Los Angeles lawyer who writes on free-speech issues, usually at the blog *The Popehat Report.*

Aha. So making your own safe space off in a corner is totally fine, but there is also this concept of a "public space," and trying to turn a public space into a safe space for one particular special interest is wrong. So what is a "public space"? It's definitely clear that a public space is *not* just "a space owned and/or run by a government"; the concept of privately owned public spaces is a well-established one. This is true even informally: it's a common moral intuition, for example, that it's less bad for a private individual to commit violations such as discriminating against races and genders than it is for, say, a shopping mall to do the same. With the /r/bitcoin subreddit, one can make the case, regardless of who technically owns the top moderator position in the subreddit, that the subreddit very much is a public space. A few arguments particularly stand out:

- It occupies "prime real estate," specifically the word "bitcoin," which makes people consider it to be the default place to discuss Bitcoin.

- The value of the space was created not just by Theymos, but by thousands of people who arrived on the subreddit to discuss Bitcoin with an implicit expectation that it is, and will continue to be, a public space for discussing Bitcoin.

- Theymos's shift in policy was a surprise to many people, and it was not foreseeable ahead of time that it would take place.

If, instead, Theymos had created a subreddit called /r/bitcoinsmallblockers, and explicitly said that it was a curated space for small-block and attempting to instigate controversial hard forks was not welcome, then it seems likely that very few people would have seen anything

wrong about this. They would have opposed his ideology, but few (at least in blockchain communities) would try to claim that it's *improper* for people with ideologies opposed to their own to have spaces for internal discussion. But back in reality, Theymos tried to "annex a public space and demand that people within the space conform to his private norms," and so we have the Bitcoin community block-size schism, a highly acrimonious fork and chain split, and now a cold peace between Bitcoin and Bitcoin Cash.*

DEPLATFORMING

About a year ago at Deconomy** I publicly shouted down Craig Wright, a scammer claiming to be Satoshi Nakamoto, finishing my explanation of why the things he says make no sense with the question "Why is this fraud allowed to speak at this conference?"

---

* Bitcoin Cash is a fork of Bitcoin created in 2017 to increase the system's ability to handle large transaction volumes and serve as a medium of exchange.
** A Korean conference "striving to develop the concept of distributed economy" that was held in 2018 and 2019.

Of course, Craig Wright's partisans replied back with . . . accusations of censorship.

Did I try to "silence" Craig Wright? I would argue, no. One could argue that this is because "Deconomy is not a public space," but I think the much better argument is that a conference is fundamentally different from an internet forum. An internet forum can actually try to be a fully neutral medium for discussion where anything goes; a conference, on the other hand, is by its very nature a highly curated list of presentations, allocating a limited number of speaking slots and actively channeling a large amount of attention to those lucky enough to get a chance to speak. A conference is an editorial act by the organizers, saying "here are some ideas and views that we think people really should be exposed to." Every conference "censors" almost every viewpoint because there's not enough space to give them all a chance to speak, and this is inherent in the format; so raising an objection to a conference's judgment in making its selections is absolutely a legitimate act.

This extends to other kinds of selective platforms. Online platforms such as Facebook, Twitter, and YouTube already engage in active selection through algorithms that influence what people are more likely to be recommended. Typically, they do this for selfish reasons, setting up their algorithms to maximize "engagement" with their platform, often with unintended byproducts like promoting flat earth conspiracy theories. So given that these platforms are already engaging in (automated) selective presentation, it seems eminently reasonable to criticize them for not directing these same levers toward more pro-social objectives, or at the least pro-social objectives that all major reasonable political tribes agree on (e.g., quality intellectual discourse). Additionally, the "censorship" doesn't seriously block anyone's ability to learn Craig Wright's side of the story; you can just go visit his website, here you go: coingeek.com. **If someone is already operating a plat-**

form that makes editorial decisions, asking them to make such decisions with the same magnitude but with more pro-social criteria seems like a very reasonable thing to do.

A more recent example of this principle at work is the #DelistBSV campaign, where some cryptocurrency exchanges, most famously Binance, removed support for trading BSV (the Bitcoin fork promoted by Craig Wright). Once again, many people, even reasonable people, accused this campaign of being an exercise in censorship, raising parallels to credit card companies blocking WikiLeaks:

Angela Walch
@angela_walch

Following

What this phenomenon suggests is that the #crypto community's commitment to 'censorship-resistance' and getting rid of human agency/discretion may be about having the power to make the decisions to censor or not.

Power transfer, rather than power distribution.

3:43 PM - 15 Apr 2019

8 Retweets 39 Likes

I personally have been a critic of the power wielded by centralized exchanges. Should I oppose #DelistBSV on free-speech grounds? I would argue, no; it's okay to support it, but this is definitely a much closer call.

Many #DelistBSV participants like Kraken are definitely not "anything goes" platforms; they already make many editorial decisions about which currencies they accept and refuse. Kraken only accepts about a dozen currencies, so they are passively "censoring" almost everyone. Shapeshift supports more currencies but it does not support SPANK, or even KNC. So in these two cases, delisting BSV is more like reallocation of a scarce resource (attention/legitimacy) than it is like censorship. Binance is a bit different; it does accept a very large array of cryptocurrencies, adopting a philosophy much closer to "anything goes," and it does have a unique position as market leader with a lot of liquidity.

That said, one can argue two things in Binance's favor. First of all, censorship is retaliating against a truly malicious exercise of censorship on the part of core BSV community members when they threatened critics like Peter McCormack with legal letters; in "anarchic" environments with large disagreements on what the norms are, "an eye for an eye" in-kind retaliation is one of the better social norms to have because it ensures that people only face punishments that they in some sense have through their own actions demonstrated they believe are legitimate. Furthermore, the delistings won't make it that hard for people to buy or sell BSV; Coinex has said that they will not delist (and I would actually oppose second-tier "anything goes" exchanges delisting). But the delistings do send a strong message of social condemnation of BSV, which is useful and needed. So there's a case to support all delistings so far, though on reflection Binance refusing to delist "because freedom" would have also been not as unreasonable as it seems at first glance.

It's in general absolutely reasonable to oppose the existence of a concentration of power, but support that concentration of power being used for purposes that you consider pro-social as long as that concentration exists; see Bryan Caplan's exposition on reconciling supporting open borders and also supporting anti-ebola restrictions for an example in a different field.* Opposing concentrations of power only requires that one believe those concentrations of power to be *on balance* harmful and abusive; it does not mean that one must oppose all things that those concentrations of power do.

If someone manages to make a *completely permissionless* cross-chain decentralized exchange that facilitates trade between any asset and any other asset, then being "listed" on the exchange would *not* send a social signal, because everyone is listed; and I would support such an exchange existing even if it supports trading BSV. The thing that I do support is BSV being removed from already exclusive positions that confer higher tiers of legitimacy than simple existence.

So, to conclude: censorship in public spaces is bad, even if the public spaces are non-governmental; censorship in genuinely private spaces (especially spaces that are not "defaults" for a broader community) can be okay; ostracizing projects with the goal and effect of denying access to them is bad; ostracizing projects with the goal and effect of denying them scarce legitimacy can be okay.

---

* Bryan Caplan, "Ebola and Open Borders," EconLog, October 16, 2014.

CONTROL AS LIABILITY

vitalik.ca
May 9, 2019

The regulatory and legal environment around internet-based services and applications has changed considerably over the last decade. When large-scale social-networking platforms first became popular in the 2000s, the general attitude toward mass data collection was essentially "why not?" This was the age of Mark Zuckerberg saying the age of privacy is over and Eric Schmidt arguing, "If you have something that you don't want anyone to know, maybe you shouldn't be doing it in the first place." And it made personal sense for them to argue this: every bit of data you can get about others was a potential machine-learning advantage for you, every single restriction a weakness, and if something happened to that data, the costs were relatively minor. Ten years later, things are very different.

It is especially worth zooming in on a few particular trends.

- **PRIVACY:** Over the last ten years, a number of privacy laws have been passed, most aggressively in Europe but also elsewhere—the most recent being the GDPR. The GDPR has many parts, but among the most prominent are: (i) requirements for explicit consent, (ii) requirement to have a legal basis to process data, (iii) users' right to download all their data, (iv) users' right to require you to delete all their data. Other jurisdictions are exploring similar rules.

▢ **DATA-LOCALIZATION RULES:** India, Russia, and many other jurisdictions increasingly have or are exploring rules that require data on users within the country to be stored inside the country. And even when explicit laws do not exist, there's a growing shift toward concern around data being moved to countries that are perceived to not sufficiently protect it.

▢ **SHARING-ECONOMY REGULATION:** Sharing-economy companies such as Uber are having a hard time arguing to courts that, given the extent to which their applications control and direct drivers' activity, they should not be legally classified as employers.

▢ **CRYPTOCURRENCY REGULATION:** A recent FinCEN guidance attempts to clarify what categories of cryptocurrency-related activity are and are not subject to regulatory licensing requirements in the United States. Running a hosted wallet? Regulated. Running a wallet where the user controls their funds? Not regulated. Running an anonymizing mixing service? If you're *running* it, regulated. If you're just writing code . . . *not regulated.*

The FinCEN cryptocurrency guidance is not at all haphazard; rather, it's trying to separate out categories of applications where the developer is actively controlling funds, from applications where the developer has no control. The guidance carefully separates out how *multisignature wallets*, where keys are held by both the operator and the user, are sometimes regulated and sometimes not:

> If the multiple-signature wallet provider restricts its role to creating un-hosted wallets that require adding a second authorization key to the wallet owner's private key in order to validate and complete transactions, the provider is not a money transmitter because it does not accept and transmit value. On the other hand, if . . . the value is represented as an entry in the accounts of the provider, the owner does not interact with the payment system directly, or the provider maintains total independent control of the value, the provider will also qualify as a money transmitter.

Although these events are taking place across a variety of contexts and industries, I would argue that there is a common trend at play. And the trend is this: **control over users' data and digital possessions and activity is rapidly moving from an asset to a liability**. Before, every bit of control you have was good: it gives you more flexibility to earn revenue, if not now then in the future. Now, every bit of control you have is a liability: you might be regulated because of it. If you exhibit control over your users' cryptocurrency, you are a money transmitter. If you have "sole discretion over fares, and can charge drivers a cancellation fee if they choose not to take a ride, prohibit drivers from picking up passengers not using the app, and suspend or deactivate drivers' accounts," you are an employer. If you control your users' data, you're required to make sure you can argue just cause, have a compliance officer, and give your users access to download or delete the data.

If you are an application builder, and you are both lazy and fear legal trouble, there is one easy way to make sure that you violate none of the above new rules: *don't build applications that centralize control.* If you build a wallet where the user holds their private keys, you really are still "just a software provider." If you build a "decentralized Uber" that really is just a slick UI combining a payment system, a reputation system, and a search engine, and you don't control the components yourself, you really won't get hit by many of the same legal issues. If you build a website that just . . . doesn't collect data, you don't have to even think about the GDPR.

This kind of approach is of course not realistic for everyone. There will continue to be many cases where going without the conveniences of centralized control simply sacrifices too much for both developers and users, and there are also cases where the business model considerations that mandate a more centralized approach (e.g., it's easier to prevent nonpaying users from using software if the software stays on your servers) win out. But we're definitely very far from having explored the full range of possibilities that more decentralized approaches offer.

Generally, unintended consequences of laws, discouraging entire categories of activity when one wanted only to surgically forbid a few specific things, are considered to be a bad thing. Here, though, I would argue that the forced shift in developers' mindsets, from "I want to control more things just in case" to "I want to control fewer things just in case," also has many positive consequences. Voluntarily giving up control, and voluntarily taking steps to deprive oneself of the ability to do mischief, does not come naturally to many people, and while ideologically driven decentralization-maximizing projects exist today, it's not at all obvious at first glance that such services will continue to dominate as the industry mainstreams. What this trend in regulation does, however, is to give a big nudge in favor of those applications that are willing to take the centralization-minimizing, user-sovereignty-maximizing "can't be evil" route.

Hence, even though these regulatory changes are arguably not pro-freedom, at least if one is concerned with the freedom of application developers, and the transformation of the internet into a subject of political focus is bound to have many negative knock-on effects, the particular trend of control becoming a liability is in a strange way *even more pro-cypherpunk* (even if not intentionally!) than policies of maximizing total freedom for application developers would have been. Though the present-day regulatory landscape is very far from an optimal one from the point of view of almost anyone's preferences, it has unintentionally dealt the movement for minimizing unneeded centralization and maximizing users' control of their own assets, private keys, and data a surprisingly strong hand to execute on its vision. And it would be highly beneficial to the movement to take advantage of it.

CHRISTMAS SPECIAL

---

vitalik.ca

December 24, 2019

Since it's Christmas time now, and we're theoretically supposed to be enjoying ourselves and spending time with our families instead of waging endless holy wars on Twitter, this blog post will offer games that you can play with your friends that will help you have fun *and* at the same time understand some spooky mathematical concepts!

Emin Gün Sirer @
@el33th4co/

A vignette from the IC3 Bootcamp, where people unwind, among other things, by playing "1.58 dimensional chess," a game of Vitalik's invention that's surprisingly fun.

1.58-DIMENSIONAL CHESS

This is a variant of chess where the board is set up like this:

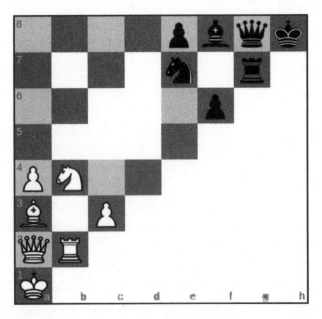

The board is still a normal eight-by-eight board, but there are only twenty-seven open squares. The other thirty-seven squares should be covered up by checkers or Go pieces or anything else to denote that they are inaccessible. The rules are the same as chess, with a few exceptions:

☐ White pawns move up, black pawns move left. White pawns take going left-and-up or right-and-up, black pawns take going left-and-down or left-and-up. White pawns promote upon reaching the top, black pawns promote upon reaching the left.

☐ No en passant, castling, or two-step-forward pawn jumps.

□ Chess pieces cannot move onto or *through* the thirty-seven covered squares. Knights cannot move onto the thirty-seven covered squares, but don't care what they move "through."

The game is called 1.58-dimensional chess because the twenty-seven open squares are chosen according to a pattern based on the Sierpinski triangle. You start off with a single open square, and then every time you double the width, you take the shape at the end of the previous step, and copy it to the top left, top right, and bottom left corners, but leave the bottom right corner inaccessible. Whereas in a one-dimensional structure, doubling the width increases the space by $2x$, and in a two-dimensional structure, doubling the width increases the space by $4x$ ($4 = 2^2$), and in a three-dimensional structure, doubling the width increases the space by $8x$ ($8 = 2^3$), here, doubling the width increases the space by $3x$ ($3 = 2^{1.58496}$), hence "1.58 dimensional."

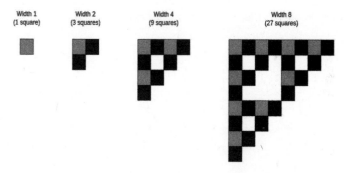

Width 1 (1 square) Width 2 (3 squares) Width 4 (9 squares) Width 8 (27 squares)

The board is constructed by starting with one square and, at each step, combining together three copies of the board from the previous step.

The game is substantially simpler and more "tractable" than full-on chess, and it's an interesting exercise in showing how in lower-dimensional spaces defense becomes much easier than offense. Note that the relative value of different pieces may change here, and new kinds of endings become possible (e.g., you can checkmate with just a bishop).

THREE-DIMENSIONAL TIC-TAC-TOE

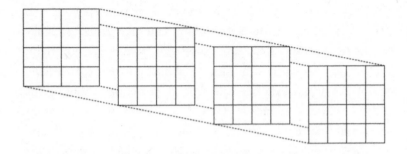

The goal here is to get four in a straight line, where the line can go in any direction, along an axis or diagonal, including between planes. For example, in this configuration, X wins:

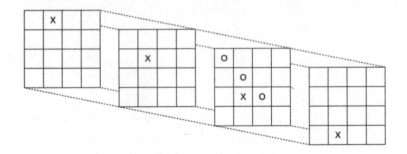

It's considerably harder than traditional two-dimensional tic-tac-toe, and hopefully much more fun!

MODULAR TIC-TAC-TOE

Here, we go back down to having two dimensions, except we allow lines to wrap around:

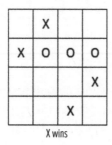

X wins

Note that we allow diagonal lines with any slope, as long as they pass through all four points. This means that lines with slope +/– 2 and +/– ½ are admissible:

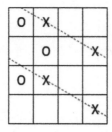

Mathematically, the board can be interpreted as a two-dimensional vector space over integers modulo 4, and the goal being to fill in a line that passes through four points over this space. Note that there exists at least one line passing through any two points.

TIC-TAC-TOE OVER THE FOUR-ELEMENT BINARY FIELD

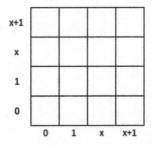

Here, we have the same concept as above, except we use an even spookier mathematical structure, the four-element field of polynomials over Z_2 modulo $x^2 + x + 1$. This structure has pretty much no reasonable geometric interpretation, so I'll just give you the addition and multiplication tables:

Addition

	0	1	x	x+1
x+1	x+1	x	1	0
x	x	x+1	0	1
1	1	0	x+1	x
0	0	1	x	x+1

Multiplication

	0	1	x	x+1
x+1	0	x+1	1	x
x	0	x	x+1	1
1	0	1	x	x+1
0	0	0	0	0

Okay, fine, here are all possible lines, excluding the horizontal and the vertical lines (which are also admissible) for brevity:

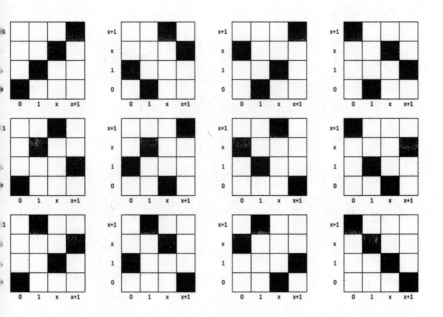

The lack of geometric interpretation does make the game harder to play; you pretty much have to memorize the twenty winning combinations, though note that they are *basically* rotations and reflections of the same four basic shapes (axial line, diagonal line, diagonal line starting in the middle, that weird thing that doesn't look like a line).

NOW PLAY 1.77-DIMENSIONAL CONNECT FOUR. I DARE YOU.

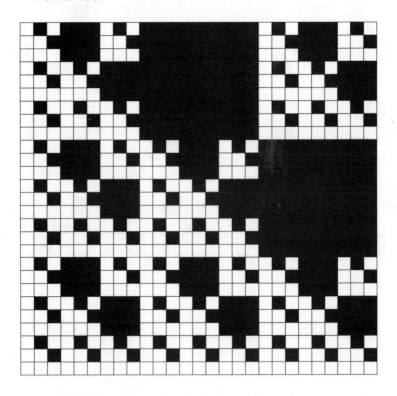

MODULAR POKER

Everyone is dealt five (you can use whatever variant poker rules you want here in terms of how these cards are dealt and whether or not players have the right to swap cards out). The cards are interpreted as: jack = 11, queen = 12, king = 0, ace = 1. A hand is stronger than another hand if it contains a longer sequence, with any constant difference between consecutive cards (allowing wraparound), than the other hand.

Mathematically, a hand is stronger if the player can come up with a line $L(x) = mx + b$ such that they have cards for the numbers $L(0), L(1) \ldots L(k)$ for the highest k.

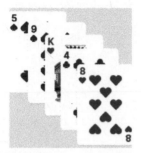

Example of a full five-card winning hand. y = 4x + 5.

To break ties between equal maximum-length sequences, count the number of distinct length-three sequences they have; the hand with more distinct length-three sequences wins.

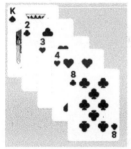

This hand has four length-three sequences:
K-2-4, K-4-8, 2-3-4, 3-8-K. This is rare.

Only consider lines of length three or higher. If a hand has three or more of the same denomination, that counts as a sequence, but if a hand has two of the same denomination, any sequences passing through that denomination only count as one sequence.

This hand has no length-three sequences.

If two hands are completely tied, the hand with the higher highest card (using J = 11, Q = 12, K = 0, A = 1 as above) wins.

Enjoy!

PART 3: PROOF OF STAKE

By the start of 2020, Ethereum had emerged from its earlier growing pains. The major security debacles had been addressed, ETH was gaining value, and Ethereum was about to become the operating system for an explosion of NFT-minted art during the COVID-19 pandemic's roving lockdowns. Distancing himself from his charismatic role during The DAO hack, Buterin stressed the principle of "credible neutrality" and reflected on how a decentralized system can achieve widespread legitimacy. He focused less on immediate crises than on the long-term problem of "public goods": How would systems based on economic incentives produce what is essential but not always profitable? Who will pay for the roads and bridges of this new world? As quickly as questions came up, so did fits and starts at answers.

The idea of the decentralized autonomous organization, or DAO, was finally becoming real. DAOs, sometimes detached entirely from any terrestrial company or foundation, were producing products and paying workers. Some were managing multimillion-dollar treasuries, others blew up as spectacular failures. By necessity, crypto communities were experimenting with new kinds of governance and decision-making processes—voting systems that balanced the power of tokens and people, identity systems based on relationships among users rather than their relationship to the state. Measures of inequality, Buterin argues, need to be rethought for a world of many overlapping forms of value. As he places a prediction-market bet on the 2020 US *presidential election, we see how confounding it can be to use software built on the protocol he designed.*

The forest can get lost in the trees. What is the point of betting in prediction markets, really? Buterin hopes that better mechanisms will make better use of the information and judgment spread out unevenly among us, guiding us toward better collective decisions. But good intentions and clever designs only go so far. The ungovernable alchemy of token prices always threatens to eclipse anything else.

Here, the transition to Ethereum 2.0—which Buterin had been hoping for since the start—is happening. By 2021, people could stake their ETH in proof of stake, even while proof of work continued a little longer. The energy waste was almost over. "Layer 2" protocols with names like "optimistic rollups" and "ZK rollups" were poised to end the delays and transaction costs that wore on anyone trying to use Ethereum for purchases or apps. Meanwhile, newer blockchains were claiming to have solved these problems from the outset, and they began taking market share from Ethereum. In the essay on "Crypto Cities," Buterin seems to have come full circle, back to the hopeful litanies on emerging projects that he used to do in Bitcoin Magazine. *But the implication is different now. Rather than supplanting old institutions like governments, blockchains are entering into relationships with them.*

Buterin has said that he learned to detest centralized platforms after the company behind the game World of Warcraft *weakened with a change to the software. ("I cried myself to sleep," he added, and gave up the game.) But he begins the final essay here by suggesting that crypto has something to learn from a concept in* Warcraft, *the "soulbound": things a player has that can't be bought or sold. Rather than relying solely on economics, on what can be bought and sold, blockchains must be able to see more clearly the humans using them. In how we design our social infrastructure our humanity is at stake.*

—NS

CREDIBLE NEUTRALITY AS A GUIDING PRINCIPLE

Nakamoto
January 3, 2020

Consider the following:

☐ People are sometimes upset at governments spending 5% of GDP to support specific public projects or specific industries, but those same people are often not upset at that same government causing much larger reallocations of capital by enforcing property rights.

☐ People are sometimes upset at blockchain projects that directly allocate (or "premine") many coins into the hands of recipients hand-picked by developers, but those same people are often not upset at the billions of dollars of value printed by major blockchains like Bitcoin and Ethereum into the hands of proof-of-work miners.

☐ People are sometimes upset at social media platforms censoring or deprioritizing content with specific disfavored political ideologies, even ideologies that the people upset at the censorship themselves disagree with, but those same people are often not upset at the fact that ride-sharing platforms kick drivers off the platform if their ratings are too low.

One possible reaction to some of these situations is to shout, "Gotcha!" and bask in the glory of having seemingly unmasked a hypocrite. And indeed, sometimes this reaction is correct. In my view, it is genuinely a mistake to treat carbon taxes as statist interventionism while treating government enforcement of property rights as just enforcement of natural law. It is genuinely also a mistake to treat miners working to secure a blockchain as laborers doing Real Thermodynamic Work worthy of compensation, but treat any attempt to compensate developers improving the blockchain's code as being an act of "printing free money."

But even if attempts to *systematize* one's intuitions often go astray, deep moral intuitions like these are rarely entirely devoid of value. And in this case, I would argue that there is a very important principle that is at play, and one that is likely to become key to the discourse of how to build efficient, pro-freedom, fair and inclusive institutions that influence and govern different spheres of our lives. And that principle is this: **when building mechanisms that decide high-stakes outcomes, it's very important for those mechanisms to be *credibly* neutral**.

MECHANISMS ARE ALGORITHMS PLUS INCENTIVES

First, what is a mechanism? Here I use the term in a way similar to that used in the game theory literature when talking about mechanism design: essentially, a mechanism is an algorithm plus incentives. A mechanism is a tool that takes in inputs from multiple people, and uses these inputs as a way to determine things about its participants' values, so as to make some kind of decision that people care about. In a well-functioning mechanism, the decision made by the mechanism is both efficient—in the sense that the decision is the best possible outcome given the participants' preferences—and incentive-compatible, meaning that people have the incentive to participate "honestly."

It's easy to come up with examples of mechanisms. A few examples:

□ **PRIVATE PROPERTY AND TRADE:** The "inputs" are users' ability to reassign ownership through donation or trade; and the "output" is a (sometimes formalized, sometimes only implied) database of who has the right to determine how each physical object is used. The goal is to encourage production of useful physical objects and put them into the hands of people who make best use of them.

□ **AUCTIONS:** The input is bids; the output is who gets the item being sold, and how much the buyer must pay.

□ **DEMOCRACY:** The input is votes; the output is who controls each seat in the government that was up for election.

□ **UPVOTES, DOWNVOTES, LIKES, AND RETWEETS ON SOCIAL MEDIA:** The input is upvotes, downvotes, likes, and retweets; the output is who sees what content. A game theory pedant may say that this is only an algorithm, not a mechanism, because it lacks built-in incentives; but future versions may well have built-in incentives.

□ **BLOCKCHAIN-AWARDED INCENTIVES FOR PROOF OF WORK AND PROOF OF STAKE:** The input is what blocks and other messages participants produce; the output is which chain the network accepts as canonical, and rewards are used to encourage "correct" behavior.

We are entering a hyper-networked, hyper-intermediated, and rapidly evolving information age, in which centralized institutions are losing public trust and people are searching for alternatives. As such, different forms of mechanisms—as a way of intelligently aggregating the wisdom of the crowds (and sifting it apart from the also ever-present non-wisdom of the crowds)—are likely to only grow more and more relevant to how we interact.

WHAT IS CREDIBLE NEUTRALITY?

Now, let us talk about this all-important idea, "credible neutrality." Essentially, a mechanism is credibly neutral if just by looking at the mechanism's design, it is easy to see that the mechanism does not discriminate for or against any specific people. The mechanism treats everyone fairly, to the extent that it's possible to treat people fairly in a world where everyone's capabilities and needs are so different. "Anyone who mines a block gets 2 ETH" is credibly neutral; "Bob gets 1,000 coins because we know he's written a lot of code and we should reward him" is not. "Any post that five people flag as being bad does not get shown" is credibly neutral; "any post that our moderation team decides is prejudiced against blue-eyed people does not get shown" is not. "The government grants a twenty-year limited monopoly to any invention" is credibly neutral (though there are serious challenges around the edges in determining what inventions qualify); "the government decides that curing cancer is important, and so appoints a committee to distribute $1 billion among people trying to cure cancer" is not.

Of course, neutrality is never total. Block rewards discriminate in favor of those that have special connections that give them access to hardware and cheap electricity. Capitalism discriminates in favor of concentrated interests and the wealthy, and against the poor and those who rely heavily on public goods. Political discourse discriminates against anything caught on the wrong side of social-desirability bias. And any mechanism that corrects for coordination failures has to make some assumptions about what those failures are, and discriminates against those whose coordination failures it underestimates. But this does not detract from the fact that some mechanisms are much more neutral than others.

This is why private property is as effective as it is: not because it is a god-given right, but because it's a credibly neutral mechanism that solves a lot of problems in society—far from all problems, but still a lot. This is why filtering by popularity is okay, but filtering by political ideology is problematic: it's easier to agree that a neutral mechanism treats everyone reasonably fairly than it is to convince a diverse group of people that some particular blacklist of unallowed political viewpoints is correct. And this is why on-chain developer rewards are viewed more suspiciously than on-chain mining rewards: it's easier to verify who's a miner than it is to verify who's a developer, and most attempts to identify who is a developer in practice easily fall prey to accusations of favoritism.

Note that it is not just neutrality that is required here, it is *credible* neutrality. That is, it is not just enough for a mechanism to not be designed to favor specific people or outcomes over others; it's also crucially important for a mechanism to be able to convince a large and diverse group of people that the mechanism at least makes that basic effort to be fair. Mechanisms such as blockchains, political systems, and social media are designed to facilitate cooperation across large, and diverse, groups of people. In order for a mechanism to actually be able to serve as this kind of common substrate, everyone participating must be able to see that the mechanism is fair, and everyone participating must be able to see that everyone else is able to see that the mechanism is fair, because everyone participating wants to be sure that everyone else will not abandon the mechanism the next day.

That is, what we need is something like a game-theoretic concept of common knowledge—or, in less mathematical terms, a widely shared notion of *legitimacy*. To achieve this kind of common knowledge of neutrality, the neutrality of the mechanism must be very easy to see—so easy to see that even a relatively uneducated observer can see it, even in the face of a hostile propaganda effort to make the mechanism seem biased and untrustworthy.

BUILDING CREDIBLY NEUTRAL MECHANISMS

There are four primary rules to building a credibly neutral mechanism:

1. Don't write specific people or specific outcomes into the mechanism

2. Open source and publicly verifiable execution

3. Keep it simple

4. Don't change it too often

Rule (1) is simple to understand. To go back to our previous examples, "Anyone who mines a block gets 2 ETH" is credibly neutral; "Bob gets 1,000 coins" is not. "Downvotes mean a post gets shown less" is credibly neutral; "prejudice against blue-eyed people means a post gets shown less" is not. "Bob" is a specific person, and "prejudice against blue-eyed people" is a specific outcome. Now, of course, Bob may genuinely be a great developer who was really valuable to some blockchain project's success and deserves a reward, and anti-blue-eyed prejudice is certainly an idea I, and hopefully you, don't want to see become prominent. But in credibly neutral mechanism design, the goal is that these desired outcomes are not written into the mechanism; instead, they are emergently discovered from the participants' actions. In a free market, the fact that Charlie's widgets are not useful but David's widgets are useful is emergently discovered through the price mechanism: eventually, people stop buying Charlie's widgets, so he goes bankrupt, while David earns a profit and can expand and make even more widgets. **Most bits of information in the output should come from the participants' inputs, not from hard-coded rules inside of the mechanism itself.**

Rule (2) is also simple to understand: the rules of the mechanism should be public, and it should be possible to publicly verify that the rules are being executed correctly. Note that in many cases, you don't want the inputs or outputs to be public; my article "On Collusion" goes into the reasons why a very strong level of privacy, where you cannot even prove how you participated if you want to, is often a good idea. Fortunately, verifiability and privacy can be achieved at the same time with a combination of zero-knowledge proofs and blockchains.

Rule (3), the idea of simplicity, is ironically the least simple. The simpler a mechanism is, and the fewer parameters a mechanism has, the less space there is to insert hidden privilege for or against a targeted group. If a mechanism has fifty parameters that interact in complicated ways, then it's likely that for any desired outcome you can find parameters that will achieve that outcome. But if a mechanism has only one or two parameters, this is much more difficult. You can create privilege for very broad groups ("demagogues," "the rich," etc.), but you cannot target a narrow group of people, and your ability to target specific outcomes goes down further with time, as there is more and more of a "veil of ignorance" between you at time A that is creating the mechanism and your beneficiaries at time B and the specific situation they will be in that might let them disproportionately benefit from the mechanism.

And this brings us to rule (4), not changing the mechanism too often. Changing the mechanism is a type of complexity, and it also "resets the clock" on the veil of ignorance, giving you the opportunity to adjust the mechanism to favor your particular friends and attack your particular enemies with the most up-to-date information about what unique positions these groups are in and how different adjustments to the mechanism would affect them.

NOT JUST NEUTRALITY: EFFICACY ALSO MATTERS

A common fallacy of the more extreme versions of the ideologies that I alluded to at the beginning of this post is a kind of neutrality maximalism: if it can't be done completely neutrally, it should not be done at all! The fallacy here is that this viewpoint achieves narrow-sense neutrality at the cost of broad-sense neutrality. For example, you can guarantee that every miner will be on the same footing as every other miner (12.5 BTC or 2 ETH per block), and that every developer will be on the same footing as every other developer (with no remuneration beyond thanks for their public service), but what you sacrifice is that development becomes highly under-incentivized relative to mining. It is unlikely that the last 20% of miners contribute more to a blockchain's success than its developers, and yet that's what the current reward structures seem to imply.

Speaking more broadly, there are many kinds of things in society that need to be produced: private goods, public goods, accurate information, good governance decisions, goods we don't value now but will value in the future, and so forth; the list goes on. Some of these things are easier to create credibly neutral mechanisms for than others. And if we adopt an uncompromising narrow-sense neutrality purism that says that only extremely credibly neutral mechanisms are acceptable, then only those problems for which such mechanisms are easy to create will be solved. The community's other needs will see no systematic support at all, and so broad-sense neutrality suffers.

Hence, the principle of credible neutrality must also be augmented with another idea, the principle of efficacy. A good mechanism is also a mechanism that actually does solve the problems that we care about. Often, this means that developers of even the most obviously credibly neutral mechanisms should be open to critique, as it's very possible for a mechanism to be both credibly neutral and terrible (as patents are often argued to be).

Sometimes, this even means that if a credibly neutral mechanism to solve some problem has not yet been found, an imperfectly neutral mechanism should be adopted in the short term. Premines and time-limited developer rewards in blockchains are one example of this; using centralized methods for detecting accounts that represent a unique human and filtering out others when decentralized methods are not yet available is another. But recognizing credible neutrality as something that is very valuable, and striving to get closer to that ideal over time, is nevertheless important.

If one is truly concerned about an imperfectly neutral mechanism leading to loss of trust or political capture, then there are ways to adopt a "fail-safe" approach to implementing it. For example, one can direct transaction fees and not issuance toward developer funding, creating a "Schelling fence"* limiting how much funding can be made. One can add time limits, or an "ice age," where the rewards fade away over time and must be renewed explicitly. One can implement the mechanism inside of a "layer 2" system,** such as

---

* This is a tweak on the concept of the "Schelling point," named for the Cold War game theorist Thomas Schelling, by the earlier-noted California psychiatrist Scott Alexander. The fence refers to a constraint on a system commonly agreed on by its participants.

** In this sense, "layer 2" refers to the infrastructures being built on top of the "layer 1" Ethereum blockchain, enabling more efficient processes for applications.

a rollup or an eth2 execution environment, that has some network effect lock-in, but can be abandoned with coordinated effort if the mechanism goes astray. When we foresee a possible breakdown in voice, we can mitigate the risks by improving freedom of exit.

Credibly neutral mechanisms for solving many kinds of problems do exist in theory, and need to be developed and improved in practice. Examples include:

☐ Prediction markets—e.g., electionbettingodds.com as a "credibly neutral" source of probabilities of who will win near-future elections

☐ Quadratic voting and funding as a way of coming to agreement on matters of governance and public goods

☐ Harberger taxes* as a more efficient alternative to pure property rights for allocating non-fungible and illiquid assets

☐ Peer prediction**

☐ Reputation systems involving transitive trust graphs

We do not yet know well what versions of ideas like these, and completely new ones, will work well, and we will need many rounds of experimentation to figure out what kinds of rules lead to good outcomes in different contexts. The need to have

---

* This is a taxation system in which people pay taxes on an asset at the rate for which they are prepared to sell it. Like the quadratic models in the previous bullet point, Harberger taxes came to the attention of Buterin and the crypto world through the book by Eric Posner and E. Glen Weyl *Radical Markets: Uprooting Capitalism and Democracy for a Just Society* (Princeton University Press, 2018).

** Peer prediction compares various user-generated ratings in a rating system and rewards users who accurately predict others' ratings. This is similar to the Schelling point concept noted above. Whereas the reputation systems in the next point depend on trust associated with particular users in a social network, peer prediction evaluates the ratings themselves relative to each other.

the mechanism's rules be open, but at the same time resistant to attack, will be a particular challenge, though cryptographic developments that allow open rules and verifiable execution and outputs together with private inputs will make some things considerably easier.

We know in principle that it is completely possible to make such robust sets of rules—as mentioned above, we've basically done it in many cases already. But as the number of software-intermediated marketplaces of different forms that we rely on keeps increasing, it becomes ever more important to make sure that these systems do not end up giving power to a select few—whether the operators of those platforms or even more powerful forces that end up capturing them—and instead create credible systems of rules that we can all get behind.

COORDINATION, GOOD AND BAD

September 11, 2021

Coordination, the ability for large groups of actors to work together for their common interest, is one of the most powerful forces in the universe. It is the difference between a king comfortably ruling a country as an oppressive dictatorship, and the people coming together and overthrowing him. It is the difference between the global temperature going up thirty-five degrees Celsius and the temperature going up by a much smaller amount if we work together to stop it. And it is the factor that makes companies, countries, and any social organization larger than a few people possible at all.

Coordination can be improved in many ways: faster spread of information, better norms that identify what behaviors are classified as cheating along with more effective punishments, stronger and more powerful organizations, tools like smart contracts that allow interactions with reduced levels of trust, governance technologies (voting, shares, decision markets . . .), and much more. And indeed, we as a species are getting better at all of these things with each passing decade.

201

But there is also a very philosophically counterintuitive dark side to coordination. **While it is emphatically true that "everyone coordinating with everyone" leads to much better outcomes than "every man for himself," what that does NOT imply is that each individual step toward more coordination is necessarily beneficial.** If coordination is improved in an unbalanced way, the results can easily be harmful.

We can think about this visually as a map, though in reality the map has many billions of "dimensions" rather than two:

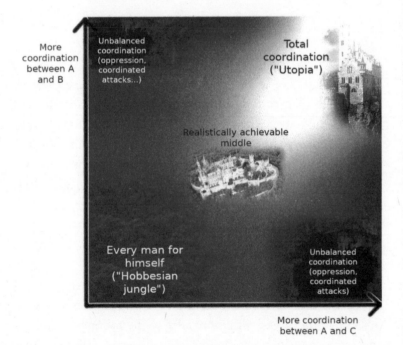

The bottom-left corner, "every man for himself," is where we don't want to be. The top-right corner, total coordination, is ideal, but likely unachievable. But the landscape in the middle is far from an even slope up, with many reasonably safe and productive places that it might be best to settle down in and many deep dark caves to avoid.

Now what are these dangerous forms of partial coordination, where someone coordinating with *some* fellow humans but not *others* leads to a deep dark hole? It's best to describe them by giving examples:

- ☐ Citizens of a nation valiantly sacrificing themselves for the greater good of their country in a war . . . when that country turns out to be World War II–era Germany or Japan

- ☐ A lobbyist giving a politician a bribe in exchange for that politician adopting the lobbyist's preferred policies

- ☐ Someone selling their vote in an election

- ☐ All sellers of a product in a market colluding to raise their prices at the same time

- ☐ Large miners of a blockchain colluding to launch a 51% attack

In all of the above cases, we see a group of people coming together and cooperating with each other, but to the great detriment of some group that is outside the circle of coordination, and thus to the net detriment of the world as a whole. In the first case, it's all the people that were the victims of the aforementioned nations' aggression that are outside the circle of coordination and suffer heavily as a result; in the second and third cases, it's the people affected by the decisions that the corrupted voter and politician are making; in the fourth case, it's the customers; and in the fifth case, it's the non-participating miners and the blockchain's users. It's not an individual defecting against the group, it's a group defecting against a broader group, often the world as a whole.

This type of partial coordination is often called "collusion," but it's important to note that the range of behaviors that we are talking about is quite broad. In normal speech, the word "collusion" tends to be used more often to describe relatively symmetrical relationships, but in the above cases there are plenty of examples with a strong asymmetric character. Even *extortionate* relationships ("vote for my preferred policies or I'll publicly reveal your affair") are a form of collusion in this sense. In the rest of this post, we'll use "collusion" to refer to "undesired coordination" generally.

EVALUATE INTENTIONS, NOT ACTIONS (!!)

One important property of especially the milder cases of collusion is that one cannot determine whether or not an action is part of an undesired collusion just by looking at the action itself. The reason is that the actions that a person takes are a combination of that person's internal knowledge, goals, and preferences together with externally imposed incentives on that person, and so the actions that people take when colluding, versus the actions that people take on their own volition (or coordinating in benign ways) often overlap.

For example, consider the case of collusion between sellers (a type of antitrust violation). If operating independently, each of three sellers might set a price for some product between $5 and $10; the differences within the range reflect difficult-to-see factors such as the seller's internal costs, their own willingness to work at different wages, supply-chain issues, and the like. But if the sellers collude, they might set a price between $8 and $13. Once again, the range reflects different possibilities regarding internal costs and other difficult-to-see factors. If you see someone selling that product for $8.75, are they doing something wrong? Without knowing whether or not they coordinated with other sellers, you can't tell! Making a law that says that selling that product for more than $8 would be a bad idea; maybe there are legitimate reasons why prices have to be high at the current time. But making a law against collusion, and successfully enforcing it, gives the ideal outcome—you get the $8.75 price if the price has to be that high to cover sellers' costs, but you don't get that price if the factors driving prices up naturally are low.

This applies in the bribery and vote-selling cases too: It may well be the case that some people vote for the Orange Party legitimately, but others vote for the Orange Party because they were paid to. From the point of view of someone determining the rules for the voting mechanism, they don't know ahead of time whether the Orange Party is good or bad. But what they do know is that a vote where people vote based on their honest internal feelings works reasonably well, but a vote where voters can freely buy and sell their votes works terribly. This is because vote selling is a tragedy of the commons: each voter only gains a small portion of the benefit from voting correctly, but would gain the full bribe if they vote the way the briber wants, and so the required bribe to lure each individual voter is far smaller than the bribe that would actually compensate the population for the costs of whatever policy the briber wants. Hence, a situation where vote selling is permitted quickly collapses into plutocracy.

DECENTRALIZATION AS ANTI-COLLUSION

But there is another, brighter and more actionable, conclusion from this line of thinking: if we want to create mechanisms that are stable, then we know that one important ingredient in doing so is finding ways to make it more difficult for collusions, especially large-scale collusions, to happen and to maintain themselves. In the case of voting, we have the secret ballot—a mechanism that ensures that voters have no way to prove to third parties how they voted, even if they want to prove it (MACI is one project trying to use cryptography to extend secret-ballot principles to an online context). This disrupts trust between voters and bribers, heavily restricting undesired collusions. In the case of antitrust and other corporate malfeasance, we often rely on whistleblowers and even give them rewards, explicitly incentivizing participants in a harmful collusion to defect. And in the case of public infrastructure more broadly, we have that oh-so-important concept: **decentralization**.

One naïve view of why decentralization is valuable is that it's about reducing risk from single points of technical failure. In traditional "enterprise" distributed systems, this is often actually true, but in many other cases we know that this is not sufficient to explain what's going on. It's instructive here to look at blockchains. A large mining pool publicly showing how they have internally distributed their nodes and network dependencies doesn't do much to calm community members scared of mining centralization. And pictures like that one showing 90% of Bitcoin hashpower at the time being capable of showing up to the same conference panel, do quite a bit to scare people:

But why is this image scary? From a "decentralization as fault tolerance" view, large miners being able to talk to each other causes no harm. But if we look at "decentralization" as being the presence of barriers to harmful collusion, then the picture becomes quite scary, because it shows that those barriers are not nearly as strong as we thought. Now, in reality, the barriers are still far from zero; the fact that those miners can easily perform technical coordination and likely are all in the same WeChat groups does *not*, in fact, mean that Bitcoin is "in practice little better than a centralized company."

So what are the remaining barriers to collusion? Some major ones include:

- □ **MORAL BARRIERS:** In *Liars and Outliers*, Bruce Schneier reminds us that many "security systems" (locks on doors, warning signs reminding people of punishments . . .) also serve a moral function, reminding potential misbehavers that they are about to conduct a serious transgression and, if they want to be a good person, they should not do that. Decentralization arguably serves that function.

- □ **INTERNAL NEGOTIATION FAILURE:** The individual companies may start demanding concessions in exchange for participating in the collusion, and this could lead to negotiation stalling outright (see "holdout problems" in economics).

- □ **COUNTER-COORDINATION:** The fact that a system is decentralized makes it easy for participants not participating in the collusion to make a fork that strips out the colluding attackers and continue the system from there. Barriers for users to join the fork are low, and the *intention* of decentralization creates moral pressure in favor of participating in the fork.

☐ **RISK OF DEFECTION:** It still is much harder for five companies to join together to do something widely considered to be bad than it is for them to join together for a non-controversial or benign purpose. The five companies do not know each other too well, so there is a risk that one of them will refuse to participate and blow the whistle quickly, and the participants have a hard time judging the risk. Individual employees within the companies may blow the whistle too.

Taken together, these barriers are substantial indeed—often substantial enough to stop potential attacks in their tracks, even when those five companies are perfectly capable of quickly coordinating to do something legitimate. Ethereum blockchain miners, for example, are perfectly capable of coordinating increases to the gas limit, but that does not mean that they can so easily collude to attack the chain.

The blockchain experience shows how designing protocols as institutionally decentralized architectures, even when it's well-known ahead of time that the bulk of the activity will be dominated by a few companies, can often be a very valuable thing. This idea is not limited to blockchains; it can be applied in other contexts as well.

FORKING AS COUNTER-COORDINATION

But we cannot always effectively prevent harmful collusions from taking place. And to handle those cases where a harmful collusion does take place, it would be nice to make systems that are more robust against them—more expensive for those colluding, and easier to recover for the system.

There are two core operating principles that we can use to

achieve this end: (1) **supporting counter-coordination** and (2) **skin in the game**. The idea behind counter-coordination is this: we know that we cannot design systems to be *passively* robust to collusions, in large part because there is an extremely large number of ways to organize a collusion and there is no passive mechanism that can detect them, but what we can do is *actively* respond to collusions and strike back.

In digital systems such as blockchains (this could also be applied to more mainstream systems—e.g., DNS),* a major and crucially important form of counter-coordination is **forking**.

If a system gets taken over by a harmful coalition, the dissidents can come together and create an alternative version of the system, which has (mostly) the same rules except that it removes the power of the attacking coalition to control the system. Forking is very easy in an open-source software context; the main challenge

---

* The domain name system is one component of the internet, which is otherwise quite decentralized, that is centralized. The early blockchain project Namecoin sought to provide a decentralized replacement. The Ethereum Name Service does this within the Ethereum ecosystem, using domains that end in .eth.

in creating a successful fork is usually gathering the **legitimacy** (game-theoretically viewed as a form of "common knowledge") needed to get all those who disagree with the main coalition's direction to follow along with you.

MARKETS AND SKIN IN THE GAME

Another class of collusion-resistance strategy is the idea of **skin in the game**. Skin in the game, in this context, basically means any mechanism that holds individual contributors in a decision individually accountable for their contributions. If a group makes a bad decision, those who approved the decision must suffer more than those who attempted to dissent. This avoids the "tragedy of the commons" inherent in voting systems.

Forking is a powerful form of counter-coordination precisely because it introduces skin in the game.

Markets are in general very powerful tools precisely because they maximize skin in the game. **Decision markets** (prediction markets used to guide decisions; also called futarchy) are an attempt to extend this benefit of markets to organizational decision-making. That said, decision markets can only solve some problems; in particular, they cannot tell us what variables we should be optimizing for in the first place.

STRUCTURING COORDINATION

This all leads us to an interesting view of what it is that people building social systems do. One of the goals of building an effective social system is, in large part, determining *the structure of coordination*: Which groups of people, and in what configurations, can come together to further their group goals, and which groups cannot?

Different coordination structures, different outcomes.

Sometimes, more coordination is good: it's better when people can work together to collectively solve their problems. At other times, more coordination is dangerous: a subset of participants could coordinate to disenfranchise everyone else. And at other times, more coordination is necessary for another reason: to enable the broader community to "strike back" against a collusion attacking the system.

In all three of those cases, there are different mechanisms that can be used to achieve these ends. Of course, it is very difficult to prevent communication outright, and it is very difficult to make coordination perfect. But there are many options in between that can nevertheless have powerful effects.

Here are a few possible coordination structuring techniques:

☐ Technologies and norms that protect privacy

☐ Technological means that make it difficult to prove how you behaved (secret ballots, MACI and similar tech)

☐ Deliberate decentralization, distributing control of some mechanism to a wide group of people that are known to not be well-coordinated

☐ Decentralization in physical space, separating out different functions (or different shares of the same function) to different locations

☐ Decentralization between role-based constituencies,

separating out different functions (or different shares of the same function) to different types of participants (e.g., in a blockchain: "core developers," "miners," "coin holders," "application developers," "users")

☐ Schelling points, allowing large groups of people to quickly coordinate around a single path forward. Complex Schelling points could potentially even be implemented in code (e.g., recovery from 51% attacks can benefit from this)

☐ Speaking a common language (or alternatively, splitting control between multiple constituencies who speak different languages)

☐ Using per-person voting instead of per-coin or per-share voting to greatly increase the number of people who would need to collude to affect a decision

☐ Encouraging and relying on defectors to alert the public about upcoming collusions

None of these strategies are perfect, but they can be used in various contexts with differing levels of success. Additionally, these techniques can and should be combined with mechanism design that attempts to make harmful collusions less profitable and more risky to the extent possible; skin in the game is a very powerful tool in this regard. Which combination works best ultimately depends on your specific use case.

Special thanks to Karl Floersch and Jinglan Wang for feedback and review.

PREDICTION MARKETS: TALES FROM THE ELECTION

vitalik.ca
February 18, 2021

Trigger warning: I express some political opinions.

Prediction markets are a subject that has interested me for many years. The idea of allowing anyone in the public to make bets about future events, and using the odds at which these bets are made as a credibly neutral source of predicted probabilities of these events, is a fascinating application of mechanism design. Closely related ideas, like futarchy, have always interested me as innovative tools that could improve governance and decision-making. And as Augur and Omen, and more recently Polymarket, have shown, prediction markets are a fascinating application of blockchains (in all three cases, Ethereum) as well.

Since the 2020 US presidential election, it seems like prediction markets are finally entering the limelight, with blockchain-based markets in particular growing from near-zero in 2016 to millions of dollars of volume in 2020. As someone who is closely interested in seeing Ethereum applications cross the chasm into widespread adoption, this of course aroused my interest. At first, I was inclined to simply watch, and not participate myself: I am not an expert on US electoral politics, so why should I expect my opinion to be more correct than that of everyone else who was already trading? But in my Twitter-sphere, I saw more and more arguments from Very Smart People whom I respected arguing that the markets were in fact being irrational and I should participate and bet against them if I can. Eventually, I was convinced.

I decided to make an experiment on the blockchain that I helped to create: I bought $2,000 worth of NTRUMP (tokens that pay $1 if Trump loses) on Augur. Little did I know then that my position would eventually increase to $308,249, earning me a profit of over $56,803, and that I would make all of these remaining bets, against willing counterparties, after Trump had already lost the election. What would transpire over the next two months would prove to be a fascinating case study in social psychology, expertise, arbitrage, and the limits of market efficiency, with important ramifications to anyone who is deeply interested in the possibilities of economic institution design.

BEFORE THE ELECTION

My first bet on this election was actually not on a blockchain at all. When Kanye announced his presidential bid in July, a political theorist whom I ordinarily quite respect for his high-quality and original thinking immediately claimed on Twitter that he was confident that this would split the anti-Trump vote and lead to a Trump victory. I remember thinking at the time that this particular opinion of his was overconfident, perhaps even a result of over-internalizing the heuristic that if a viewpoint seems clever and contrarian then it is likely to be correct. So of course I offered to make a $200 bet, myself betting the boring conventional pro-Biden view, and he honorably accepted.

The election came up again on my radar in September, and this time it was the prediction markets that caught my attention. The markets gave Trump a nearly 50% chance of winning, but I saw many Very Smart People in my Twitter-sphere whom I respected pointing out that this number seemed far too high. This of course led to the familiar "efficient markets debate": if you can buy a token that gives you $1 if Trump loses for $0.52, and Trump's actual chance of losing is much higher, why wouldn't people just come in and buy the token until the price rises more? And if nobody has done this, who are you to think that you're smarter than everyone else?

Ne0liberal's Twitter thread just before Election Day does an excellent job summarizing his case against prediction markets being accurate at that time. In short, the (non-blockchain) prediction markets that most people used at least prior to 2020 have all sorts of restrictions that make it difficult for people to participate with more than a small amount of cash. As a result, if a very smart individual or a professional organization saw a probability that they believed was wrong, they would only have a very limited ability to push the price in the direction that they believe to be correct.

The most important restrictions that the paper* points out are:

☐ Low limits (well under $1,000) on how much each person can bet

☐ High fees (e.g., a 5% withdrawal fee on PredictIt)

And this is where I pushed back against ne0liberal in September: although the stodgy old-world centralized prediction markets may have low limits and high fees, the crypto markets do not! On Augur or Omen, there's no limit to how much someone can buy or sell if they think the price of some outcome token is too low or too high. And the blockchain-based prediction markets were following the same prices as PredictIt. If the markets really were overestimating Trump because high fees and low trading limits were preventing the more cool-headed traders from outbidding the overly optimistic ones, then why would blockchain-based markets, which don't have those issues, show the same prices?

The main response my Twitter friends gave to this was that blockchain-based markets are highly niche, and very few people, particularly very few people who know much about politics,

---

* Referring to a paper mentioned in the Twitter thread mentioned above: Andrew Stershic and Kritee Gujral, "Arbitrage in Political Prediction Markets," *Journal of Prediction Markets* 14, no. 1 (2020).

have easy access to cryptocurrency. That seemed plausible, but I was not too confident in that argument. And so I bet $2,000 against Trump and went no further.

THE ELECTION

Then the election happened. After an initial scare where Trump at first won more seats than we expected, Biden turned out to be the eventual winner. Whether or not the election itself validated or refuted the efficiency of prediction markets is a topic that, as far as I can tell, is quite open to interpretation. On the one hand, by a standard Bayes rule application, I should decrease my confidence in prediction markets, at least relative to Nate Silver. Prediction markets gave a 60% chance of Biden winning, Nate Silver gave a 90% chance of Biden winning. Since Biden in fact won, this is one piece of evidence that I live in a world where Nate gives the more correct answers.

But on the other hand, you can make a case that the prediction markets better estimated the margin of victory. The median of Nate's probability distribution was somewhere around 370 of 538 Electoral College votes going to Biden:

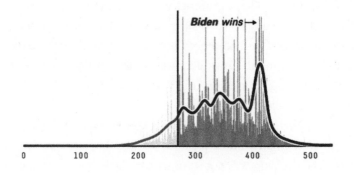

The Trump markets didn't give a probability distribution, but if you had to guess a probability distribution from the statistic "40% chance Trump will win," you would probably give one with a median somewhere around 300 Electoral College votes for Biden. The actual result: 306. So the net score for prediction markets vs. Nate seems to me, on reflection, ambiguous.

AFTER THE ELECTION

But what I could not have imagined at the time was that the election itself was just the beginning. A few days after the election, Biden was declared the winner by various major organizations and even a few foreign governments. Trump mounted various legal challenges to the election results, as was expected, and each of these challenges quickly failed. But for over a month, *the price of the NTRUMP tokens stayed at eighty-five cents!*

At the beginning, it seemed reasonable to guess that Trump had a 15% chance of overturning the results; after all, he had appointed three judges to the Supreme Court, at a time of heightened partisanship where many have come to favor team over principle. Over the next three weeks, however, it became more and more clear that the challenges were failing, and Trump's hopes continued to look grimmer with each passing day, but the NTRUMP price did not budge; in fact, it even briefly decreased to around $0.82. On December 11, more than five weeks after the election, the Supreme Court decisively and unanimously rejected Trump's attempts to overturn the vote, and the NTRUMP price finally rose . . . to $0.88.

It was in November that I was finally convinced that the market skeptics were right, and I plunged in and bet against Trump myself. The decision was not so much about the money; after all, barely two months later I would earn and donate to GiveDirectly a far larger amount simply from holding dogecoin. Rather, it was to take part in the experiment not just as an observer but as an active participant, and to improve my personal understanding of why everyone else hadn't already plunged in to buy NTRUMP tokens before me.

DIPPING IN

I bought my NTRUMP on Catnip, a front-end user interface that combines together the Augur prediction market with Balancer, a Uniswap-style constant-function market maker. Catnip was by far the easiest interface for making these trades, and in my opinion contributed significantly to Augur's usability.

There are two ways to bet against Trump with Catnip:

1. Use DAI* to buy NTRUMP on Catnip directly

2. Use Foundry to access an Augur feature that allows you to convert 1 DAI into 1 NTRUMP + 1 YTUMP + 1 ITRUMP (the "I" stands for "invalid"—more on this later), and sell the YTRUMP on Catnip

At first, I only knew about the first option. But then I discovered that Balancer has far more liquidity for YTRUMP, and so I switched to the second option.

There was also another problem: I did not have any DAI. I had ETH, and I could have sold my ETH to get DAI, but I did not want to sacrifice my ETH exposure; it would have been a shame if I earned $50,000 betting against Trump but simultaneously lost $500,000 missing out on ETH price changes. So I

---

* DAI is what is known as a "stablecoin," designed to retain a more or less constant value relative to the US dollar. It is governed by a DAO called MakerDAO.

decided to keep my ETH price exposure the same by opening up a collateralized debt position (CDP, now also called a "vault") on MakerDAO.

A CDP is how all DAI is generated: users deposit their ETH into a smart contract, and are allowed to withdraw an amount of newly-generated DAI up to two-thirds of the value of ETH that they put in. They can get their ETH back by sending back the same amount of DAI that they withdrew plus an extra interest fee (currently 3.5%). If the value of the ETH collateral that you deposited drops to less than 150% the value of the DAI you withdrew, anyone can come in and "liquidate" the vault, forcibly selling the ETH to buy back the DAI and charging you a high penalty. Hence, it's a good idea to have a high collateralization ratio in case of sudden price movements; I had over three dollars' worth of ETH in my CDP for every one dollar that I withdrew.

Recapping the above, here's the pipeline in diagram form:

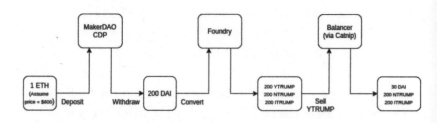

I did this many times; the slippage on Catnip meant that I could normally make trades only up to about $5,000 to $10,000 at a time without prices becoming too unfavorable (when I had skipped Foundry and bought NTRUMP with DAI directly, the limit was closer to $1,000). And after two months, I had accumulated over 367,000 NTRUMP.

WHY NOT EVERYONE ELSE?

Before I went in, I had four main hypotheses about why so few others were buying up dollars for eighty-five cents:

1. Fear that either the Augur smart contracts would break or Trump supporters would manipulate the oracle (a decentralized mechanism where holders of Augur's REP token vote by staking their tokens on one outcome or the other) to make it return a false result

2. Capital costs: to buy these tokens, you have to lock up funds for over two months, and this removes your ability to spend those funds or make other profitable trades for that duration

3. It's too technically complicated for almost everyone to trade

4. There just really are far fewer people than I thought who are actually motivated enough to take a weird opportunity even when it presents itself to them straight in the face

All four have reasonable arguments going for them. Smart contracts breaking is a real risk, and the Augur oracle had not before been tested in such a contentious environment. Capital costs are real, and while betting against something is easier in a prediction market than in a stock market, because you know that prices will never go above one dollar, locking up capital nevertheless competes with other lucrative opportunities in the crypto markets. Making transactions in dapps is technically complicated, and it's rational to have some degree of fear of the unknown.

But my experience actually going into the financial trenches, and watching the prices on this market evolve, taught me a lot about each of these hypotheses.

FEAR OF SMART-CONTRACT EXPLOITS

At first, I thought that "fear of smart-contract exploits" must have been a significant part of the explanation. But over time, I have become more convinced that it is probably *not* a dominant factor. One way to see why I think this is the case is to compare the prices for YTRUMP and ITRUMP. ITRUMP stands for "Invalid Trump"; "invalid" is an event outcome that is intended to be triggered in some exceptional cases: when the description of the event is ambiguous, when the outcome of the event is not yet known when the market is resolved, when the market is unethical (e.g., assassination markets), and a few other similar situations. In this market, the price of ITRUMP consistently stayed under $0.02. If someone wanted to earn a profit by attacking the market, it would be far more lucrative for them not to buy YTRUMP at $0.15, but instead buy ITRUMP at $0.02. If they buy a large amount of ITRUMP, they could earn a 50x return if they can force the "invalid" outcome to actually trigger. So if you fear an attack, buying ITRUMP is by far the most rational response. And yet, very few people did.

A further argument against fear of smart-contract exploits, of course, is the fact that in every crypto application *except* prediction markets (e.g., Compound, the various yield-farming schemes) people are surprisingly blasé about smart-contract risks. If people are willing to put their money into all sorts of risky and untested schemes even for a promise of mere 5% to 8% annual gains, why would they suddenly become overcautious here?

CAPITAL COSTS

Capital costs—the inconvenience and opportunity cost of locking up large amounts of money—are a challenge that I have come to appreciate much more than I did before. Just looking at the Augur side of things, I needed to lock up 308,249 DAI for an average of about two months to make a $56,803 profit. This works out to about a 175% annualized interest rate; so far, quite a good deal, even compared to the various yield-farming crazes of the summer of 2020. But this becomes worse when you take into account what I needed to do on MakerDAO. Because I wanted to keep my exposure to ETH the same, I needed to get my DAI through a CDP, and safely using a CDP required a collateral ratio of over 3x. Hence, the total amount of capital I *actually* needed to lock up was somewhere around a million dollars.

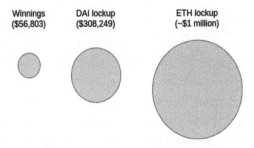

Winnings ($56,803)   DAI lockup ($308,249)   ETH lockup (~$1 million)

Now, the interest rates are looking less favorable. And if you add to that the possibility, however remote, that a smart-contract hack, or a truly unprecedented political event, actually *will* happen, it still looks less favorable.

But even still, assuming a 3x lockup and a 3% chance of Augur breaking (I had bought ITRUMP to cover the possibility that it breaks in the "invalid" direction, so I needed only worry about the risk of breaks in the "yes" direction or the funds being stolen outright), that works out to a risk-neutral rate of about 35%, and even lower once you take real human beings' views on risk into account. The deal is still very attractive, but on the other hand, it now looks very understandable that such numbers are unimpressive to people who live and breathe cryptocurrency, with its frequent 100x ups and downs.

Trump *supporters*, on the other hand, faced none of these challenges: they canceled out my $308,249 bet by throwing in a mere $60,000 (my winnings are less than this because of fees). When probabilities are close to 0 or 1, as is the case here, the game is *very* lopsided in favor of those who are trying to push the probability away from the extreme value. And this explains not just Trump; it's also the reason why all sorts of popular-among-a-niche candidates with no real chance of victory frequently get winning probabilities as high as 5%.

TECHNICAL COMPLEXITY

I had at first tried buying NTRUMP on Augur, but technical glitches in the user interface prevented me from being able to make orders on Augur directly (other people I talked to did not have this issue . . . I am still not sure what happened there). Catnip's UI is much simpler and works excellently. However, automated market makers like Balancer (and Uniswap) work best for smaller trades; for larger trades, the slippage is too high. This is a good microcosm of the broader "AMM vs. order book" debate: AMMs are more convenient but order books really do work better for large trades. Uniswap v3 is introducing an AMM design that has better capital efficiency; we shall see if that improves things.

There were other technical complexities too, though fortunately they all seem to be easily solvable. There is no reason why an interface like Catnip could not integrate the "DAI → Foundry → sell YTRUMP" path into a contract so that you could buy NTRUMP that way in a single transaction. In fact, the interface could even check the price and liquidity properties of the "DAI → NTRUMP" path and the "DAI → Foundry → sell YTRUMP" path and give you the better trade automatically. Even withdrawing DAI from a MakerDAO CDP can be included in that path. My conclusion here is optimistic: technical complexity issues were a real barrier to participation this round, but things will be much easier in future rounds as technology improves.

INTELLECTUAL UNDERCONFIDENCE

And now we have the final possibility: that many people (and smart people in particular) have a pathology that they suffer from excessive humility, and too easily conclude that if no one else has taken some action, then there must therefore be a good reason why that action is not worth taking.

Eliezer Yudkowsky spends the second half of his excellent book *Inadequate Equilibria* making this case, arguing that too many people overuse "modest epistemology," and we should be much more willing to act on the results of our reasoning, even when the result suggests that the great majority of the population is irrational or lazy or wrong about something. When I read those sections for the first time, I was unconvinced; it seemed like Eliezer was simply being overly arrogant. But having gone through this experience, I have come to see some wisdom in his position.

This was not my first time seeing the virtues of trusting one's own reasoning firsthand. When I had originally started working on Ethereum, I was at first beset by fear that there must be some very good reason the project was doomed to fail. A fully programmable smart-contract-capable blockchain, I reasoned, was clearly such a great improvement over what came before, that surely many other people must have thought of it before I did. And so I fully expected that, as soon as I published the idea, many very smart cryptographers would tell me the very good reasons why something like Ethereum was fundamentally impossible. And yet, no one ever did.

Of course, not everyone suffers from excessive modesty. Many of the people making predictions *in favor* of Trump winning the election were arguably fooled by their own excessive contrarianism. Ethereum benefited from my youthful suppression of my own modesty and fears, but there are plenty of other projects that could have benefited from more intellectual humility and avoided failures.

Not a sufferer of excessive modesty.

But nevertheless it seems to me more true than ever that, as goes the famous Yeats quote, "the best lack all conviction, while the worst are full of passionate intensity." Whatever the faults of overconfidence or contrarianism sometimes may be, it seems clear to me that spreading a society-wide message that the solution is to simply trust the existing outputs of society, whether those come in the form of academic institutions, media, governments, or markets, is not the solution. All of these institutions can only work precisely because of the presence of individuals who think that they do not work, or who think that they can be wrong some of the time.

LESSONS FOR FUTARCHY

Seeing the importance of capital costs and their interplay with risks firsthand is also important evidence for judging systems like futarchy. Futarchy, and "decision markets" more generally, is an important and potentially very socially useful application of prediction markets. There is not much social value in having slightly more accurate predictions of who will be the next president. But there is a lot of social value in having **conditional predictions**: *If we do A, what's the chance it will lead to some good thing X, and if we do B instead what are the chances then?* Conditional predictions are important because they do not just satisfy our curiosity; they can also help us make decisions.

Though electoral-prediction markets are much less useful than conditional predictions, they can help shed light on an important question: How robust are they against manipulation or even just biased and wrong opinions? We can answer this question by looking at how difficult arbitrage is: Suppose that a conditional-prediction market currently gives probabilities that (in your opinion) are *wrong* (could be because of ill-informed traders or an explicit manipulation attempt; we don't really care). How much of an impact can you have, and how much profit can you make, by setting things right?

Let's start with a concrete example. Suppose that we are trying to use a prediction market to choose between decision A and decision B, where each decision has some probability of achieving some desirable outcome. Suppose that your opinion is that decision A has a 50% chance of achieving the goal, and decision B has a 45% chance. The market, however, (in your opinion wrongly) thinks decision B has a 55% chance and decision A has a 40% chance.

Probability of good outcome if we choose strategy . . .	Current market position	Your opinion
A	40%	50%
B	55%	45%

Suppose that you are a small participant, so your individual bets won't affect the outcome; only many bettors acting together could. How much of your money should you bet?

The standard theory here relies on the Kelly criterion. Essentially, you should act to maximize the expected logarithm of your assets. In this case, we can solve the resulting equation. Suppose you invest portion r of your money into buying A-token for \$0.40. Your expected new log-wealth, from your point of view, would be:

$$0.5 \times \log((1 - r) + \frac{r}{0.4}) + 0.5 \times \log(1 - r)$$

The first term is the 50% chance (from your point of view) that the bet pays off, and the portion r that you invest grows by $2.5x$ (as you bought dollars at forty cents). The second term is the 50% chance that the bet does not pay off, and you lose the portion you bet. We can use calculus to find the r that maximizes this. The answer is $r = 1/6$. If other people buy and the price for A on the market gets up to 47% (and B gets down to 48%), we can redo the calculation for the last trader who would flip the market over to make it correctly favor A:

$$0.5 \times \log\left((1 - r) + \frac{r}{0.4}\right) + 0.5 \times \log(1 - r)$$

Here, the expected log-wealth-maximizing r is a mere 0.0566. The conclusion is clear: when decisions are close and when there is a lot of noise, it turns out that it only makes sense to invest a small portion of your money in a market. And this is assuming rationality; most people invest less into uncertain gambles than the Kelly criterion says they should. Capital costs stack on top even further. But if an attacker *really* wants to force outcome B through because they want it to happen for personal reasons, they can simply put *all* of their capital toward buying that token. All in all, the game can easily be lopsided more than twenty to one in favor of the attacker.

Of course, in reality attackers are rarely willing to stake all their funds on one decision. And futarchy is not the only mechanism that is vulnerable to attacks: stock markets are similarly vulnerable, and non-market decision mechanisms can also be manipulated by determined wealthy attackers in all sorts of ways. But nevertheless, we should be wary of assuming that futarchy will propel us to new heights of decision-making accuracy.

Interestingly enough, the math seems to suggest that futarchy would work best when the expected manipulators want to push the outcome toward an extreme value. An example of this might be liability insurance, as someone wishing to improperly obtain insurance would effectively be trying to force the market-estimated probability that an unfavorable event will happen down to zero. And as it turns out, liability insurance is futarchy inventor Robin Hanson's new favorite policy prescription.

CAN PREDICTION MARKETS BECOME BETTER?

The final question to ask is: Are prediction markets doomed to repeat errors as grave as giving Trump a 15% chance of overturning the election in early December, and a 12% chance of overturning it even after the Supreme Court, including three judges whom he appointed, told him to screw off? Or could the markets improve over time? My answer is, surprisingly, emphatically on the optimistic side, and I see a few reasons for optimism.

MARKETS AS NATURAL SELECTION

First, these events have given me a new perspective on how market efficiency and rationality might actually come about. Too often, proponents of market-efficiency theories claim that market efficiency results because most participants are rational (or at least the rationals outweigh any coherent group of deluded people), and this is true as an axiom. But, instead, we could take an *evolutionary* perspective on what is going on.

Crypto is a young ecosystem. It is an ecosystem that is still quite disconnected from the mainstream, Elon's recent tweets* notwithstanding, and that does not yet have much expertise in the minutiae of electoral politics. Those who are experts in electoral politics have a hard time getting into crypto, and crypto has a large presence of not-always-correct forms of contrarianism especially when it comes to politics. But what happened this year is that within the crypto space, prediction-market users who correctly expected Biden to win got an 18% increase to their capital, and prediction market users who incorrectly expected Trump to

---

* Referring, of course, to the billionaire Elon Musk, whose tweets about cryptocurrency have the potential to cause significant shifts in value.

win got a 100% decrease to their capital (or at least the portion they put into the bet).

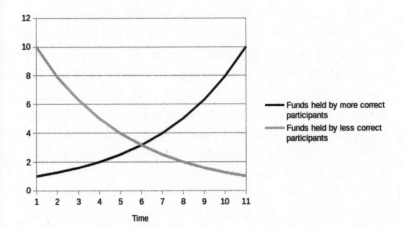

Thus, there is a *selection* pressure in favor of the type of people who make bets that turn out to be correct. After ten rounds of this, good predictors will have more capital to bet with, and bad predictors will have less capital to bet with. This does *not* rely on anyone "getting wiser" or "learning their lesson" or any other assumption about humans' capacity to reason and learn. It is simply a result of selection dynamics such that, over time, participants that are good at making correct guesses will come to dominate the ecosystem.

Note that prediction markets fare better than stock markets in this regard: the "nouveaux riches" of stock markets often arise from getting lucky on a single thousandfold gain, adding a lot of noise to the signal, but in prediction markets, prices are bounded between 0 and 1, limiting the impact of any one single event.

BETTER PARTICIPANTS AND BETTER TECHNOLOGY

Second, prediction markets themselves will improve. User interfaces have greatly improved already, and will continue to improve further. The complexity of the MakerDAO → Foundry → Catnip cycle will be abstracted away into a single transaction. Blockchain-scaling technology will improve, reducing fees for participants.

Third, the demonstration that we saw of the prediction market working correctly will ease participants' fears. Users will see that the Augur oracle is capable of giving correct outputs even in very contentious situations (this time, there were two rounds of disputes, but the no side nevertheless cleanly won). People from outside the crypto space will see that the process works and be more inclined to participate. Perhaps even Nate Silver himself might get some DAI and use Augur, Omen, Polymarket, and other markets to supplement his income in 2022 and beyond.

Fourth, prediction market tech itself could improve. Here is a proposal from myself on a market design that could make it more capital-efficient to simultaneously bet against many unlikely events, helping to prevent unlikely outcomes from getting irrationally high odds. Other ideas will surely spring up, and I look forward to seeing more experimentation in this direction.

CONCLUSION

This whole saga has proven to be an incredibly interesting direct trial-by-first-test of prediction markets and how they collide with the complexities of individual and social psychology. It shows a lot about how market efficiency actually works in practice, what are the limits of it, and what could be done to improve it.

It has also been an excellent demonstration of the power of blockchains; in fact, it is one of the Ethereum applications that have provided to me the most concrete value. Blockchains are often criticized for being speculative toys and not doing anything meaningful except for self-referential games (tokens, with yield farming, whose returns are powered by . . . the launch of other tokens). There are certainly exceptions that the critics fail to recognize; I personally have benefited from ENS and even from using ETH for payments on several occasions where all credit card options failed. But over the last few months, it seems like we have seen a rapid burst in Ethereum applications being concretely useful for people and interacting with the real world, and prediction markets are a key example of this.

I expect prediction markets to become an increasingly important Ethereum application in the years to come. The 2020 election was only the beginning; I expect more interest in prediction markets going forward, not just for elections but for conditional predictions, decision-making, and other applications as well. The amazing promises of what prediction markets could bring if they work mathematically optimally will, of course, continue to collide with the limits of human reality, and hopefully, over time, we will get a much clearer view of exactly where this new social technology can provide the most value.

Special thanks to Jeff Coleman, Karl Floersch, and Robin Hanson for critical feedback and review.

THE MOST IMPORTANT SCARCE RESOURCE IS LEGITIMACY

vitalik.ca
March 23, 2021

The Bitcoin and Ethereum blockchain ecosystems both spend far more on network security—the goal of proof-of-work mining—than they do on everything else combined. The Bitcoin blockchain has paid an average of about $38 million per day in block rewards to miners since the start of the year, plus about $5 million per day in transaction fees. The Ethereum blockchain comes in second, at $19.5 million per day in block rewards plus $18 million per day in transaction fees. Meanwhile, the Ethereum Foundation's annual budget, paying for research, protocol development, grants, and all sorts of other expenses, is a mere $30 million per year. Non-EF-sourced funding exists too, but it is at most only a few times larger. Bitcoin ecosystem expenditures on R&D are likely even lower. Bitcoin-ecosystem R&D is largely funded by companies (with $250 million total raised so far), and about fifty-seven employees; assuming fairly high salaries and the likelihood that many paid developers working for companies are not being counted, that works out to about $20 million per year.

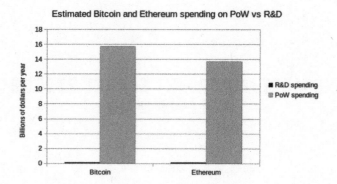

Clearly, this expenditure pattern *is a massive misallocation* of resources. The last 20% of network hashpower provides *vastly* less value to the ecosystem than those same resources would if they had gone into research and core protocol development. So why not just . . . cut the PoW budget by 20% and redirect the funds to those other things instead?

The standard answer to this puzzle has to do with concepts like "public choice theory" and "Schelling fences": Even though we could easily identify some valuable public goods to redirect some funding to as a one-off, making a *regular institutionalized* pattern of such decisions carries risks of political chaos and capture that are in the long run not worth it. But regardless of the reasons why, we are faced with this interesting fact that **the organisms that are the Bitcoin and Ethereum ecosystems are capable of summoning up billions of dollars of capital, but have strange and hard-to-understand restrictions on where that capital can go**.

The powerful social force that is creating this effect is worth understanding. As we are going to see, it's also the same social force behind why the Ethereum ecosystem is capable of summoning up these resources in the first place (and the technologically near-identical Ethereum Classic* is not). It's also a social force that is key to helping a chain recover from a 51% attack. And it's a social force that underlies all sorts of extremely powerful mechanisms far beyond the blockchain space. For reasons that will be clear in the upcoming sections, I will give this powerful social force a name: **legitimacy**.

COINS CAN BE OWNED BY SOCIAL CONTRACTS

To better understand the force that we are getting at, another important example is the epic saga of Steem and Hive. In early 2020, Justin Sun bought Steem-the-company, which is not the same thing as Steem-the-blockchain but did hold about 20% of the STEEM token supply. The community, naturally, did not trust Justin Sun. So they made an on-chain vote to formalize what they considered to be a longstanding "gentlemen's agreement" that Steem-the-company's coins were held in trust for the common good of Steem-the-blockchain and should not be used to vote. With the help of coins held by exchanges, Justin Sun made a counterattack and won control of enough delegates to unilaterally control the chain. The community saw no further in-protocol options. So instead they made a fork of Steem-the-blockchain, called Hive, and copied over all of the STEEM token balances—except those, including Justin Sun's, which participated in the attack.

---

* Ethereum Classic is the branch of the Ethereum blockchain that did not adopt the "hard fork" and erase the 2016 hack of The DAO. Before that event it is the same as Ethereum; after the event it diverges.

Hive Ecosystem

Hive has a thriving ecosystem of apps, communities & individuals, leveraging the Hive blockchain & decentralised structure.

Splinterlands ⬈   Peakd ⬈   Hive.Blog ⬈   3Speak ⬈

Brewmaster ⬈   Ecency ⬈   Rabona ⬈   D.Buzz ⬈

And they got plenty of applications on board. If they had not managed this, far more users would have either stayed on Steem or moved to some different project entirely.

The lesson that we can learn from this situation is this: *Steem-the-company never actually "owned" the coins*. If they did, they would have had the practical ability to use, enjoy, and abuse the coins in whatever way they wanted. But in reality, when the company tried to enjoy and abuse the coins in a way that the community did not like, *they were successfully stopped*. What's going on here is a pattern of a similar type to what we saw with the not-yet-issued Bitcoin and Ethereum coin rewards: the coins were ultimately owned not by a cryptographic key, but by *some kind of social contract*.

We can apply the same reasoning to many other structures in the blockchain space. Consider, for example, the ENS root multisig.* The root multisig is controlled by seven prominent ENS and Ethereum

---

* ENS is the Ethereum Name Service, the registrar for .eth domains widely used in the Ethereum ecosystem. A "root multisig" is an Ethereum wallet that controls a particular contract, in this case the contract governing the ENS system.

community members. But what would happen if four of them were to come together and "upgrade" the registrar to one that transfers all the best domains to themselves? Within the context of ENS-the-smart-contract-system, they have the complete and unchallengeable ability to do this. But if they actually tried to abuse their technical ability in this way, what would happen is clear to anyone: they would be ostracized from the community, the remaining ENS community members would make a new ENS contract that restores the original domain owners, and every Ethereum application that uses ENS would repoint their UI to use the new one.

This goes well beyond smart-contract structures. Why is it that Elon Musk can sell an NFT of Elon Musk's tweet, but Jeff Bezos would have a much harder time doing the same? Elon and Jeff have the same level of ability to screenshot Elon's tweet and stick it into an NFT dapp, so what's the difference? To anyone who has even a basic intuitive understanding of human social psychology (or the fake-art scene), the answer is obvious: Elon selling Elon's tweet is *the real thing*, and Jeff doing the same is not. Once again, millions of dollars of value are being controlled and allocated, not by individuals or cryptographic keys, but by social conceptions of legitimacy.

And, going even further out, legitimacy governs all sorts of social status games, intellectual discourse, language, property rights, political systems, and national borders. Even blockchain consensus works the same way: the only difference between a soft fork that gets accepted by the community and a 51% censorship attack, after which the community coordinates an extra-protocol recovery fork to take out the attacker, is legitimacy.

SO WHAT IS LEGITIMACY?

To understand the workings of legitimacy, we need to dig down into some game theory. There are many situations in life that demand **coordinated behavior**: if you act in a certain way alone, you are likely to get nowhere (or worse), but if everyone acts together, a desired result can be achieved.

	A	B
A	(5, 5)	(0, 0)
B	(0, 0)	(5, 5)

An abstract coordination game. You benefit heavily from making the same move as everyone else.

One natural example is driving on the left vs. the right side of the road: it doesn't really matter what side of the road people drive on, as long as they drive on the same side. If you switch sides at the same time as everyone else, and most people prefer the new arrangement, there can be a net benefit. But if you switch sides alone, no matter how much you prefer driving on the other side, the net result for you will be quite negative.

Now, we are ready to define legitimacy.

> **Legitimacy is a pattern of higher-order acceptance. An outcome in some social context is *legitimate* if the people in that social context broadly accept and play their part in enacting that outcome, and each individual person does so because they expect everyone else to do the same.**

Legitimacy is a phenomenon that arises naturally in coordination games. If you're not in a coordination game, there's no reason to act according to your expectation of how other people will act, and so legitimacy is not important. But as we have seen, coordination games are everywhere in society, and so legitimacy turns out to be quite important indeed. In almost any environment with coordination games that exists for long enough, there inevitably emerge some mechanisms by which people can choose which decision to take. These mechanisms are powered by an established culture in which everyone pays attention to these mechanisms and (usually) does what they say. Each person reasons that because *everyone else* follows these mechanisms, if they do something different they will only create conflict and suffer, or at least be left in a lonely forked ecosystem all by themselves. If a mechanism successfully has the ability to make these choices, then that mechanism has legitimacy.

In any context where there's a coordination game that has existed for long enough, there's likely a conception of legitimacy. **And blockchains are full of coordination games.** Which client software do you run? Which decentralized domain-name registry do you ask for, and which address corresponds to a .eth name? Which copy of the Uniswap contract do you accept as being "the" Uniswap exchange?* Even NFTs are a coordination

---

* Names ending with .eth are part of the Ethereum Name System, a domain registry that links a domain name with an Ethereum address. Uniswap is a token-exchange platform that operates as a smart-contract protocol on the Ethereum blockchain; it is an open-source software and can be copied and modified by anyone motivated to do so.

game. The two largest parts of an NFT's value are (i) pride in holding the NFT and ability to show off your ownership, and (ii) the possibility of selling it in the future. For both of these components, it's really, really important that whatever NFT you buy is recognized as *legitimate* by everyone else. In all of these cases, there's a great benefit to having the same answer as everyone else, and the mechanism that determines that equilibrium has a lot of power.

THEORIES OF LEGITIMACY

There are many different ways in which legitimacy can come about. In general, legitimacy arises because the thing that gains legitimacy is psychologically appealing to most people. But of course, people's psychological intuitions can be quite complex. It is impossible to make a full listing of theories of legitimacy, but we can start with a few:

- **LEGITIMACY BY BRUTE FORCE:** Someone convinces everyone that they are powerful enough to impose their will and resisting them will be very hard. This drives most people to submit because each person expects that *everyone else* will be too scared to resist as well.

- **LEGITIMACY BY CONTINUITY:** If something was legitimate at time T, it is by default legitimate at time T + 1.

- **LEGITIMACY BY FAIRNESS:** Something can become legitimate because it satisfies an intuitive notion of fairness. See also: my post on credible neutrality, though note that this is not the only kind of fairness.

- **LEGITIMACY BY PROCESS:** If a process is legitimate, the outputs of that process gain legitimacy (e.g., laws passed by democracies are sometimes described in this way).

☐ **LEGITIMACY BY PERFORMANCE:** If the outputs of a process lead to results that satisfy people, then that process can gain legitimacy (e.g., successful dictatorships are sometimes described in this way).

☐ **LEGITIMACY BY PARTICIPATION:** If people participate in choosing an outcome, they are more likely to consider it legitimate. This is similar to fairness, but not quite: it rests on a psychological desire to be consistent with your previous actions.

Note that legitimacy is a descriptive concept; something can be legitimate even if you personally think that it is horrible. That said, if enough people think that an outcome is horrible, there is a higher chance that some event will happen in the future that will cause that legitimacy to go away, often at first gradually, then suddenly.

LEGITIMACY IS A POWERFUL SOCIAL TECHNOLOGY, AND WE SHOULD USE IT

The public-goods funding situation in cryptocurrency ecosystems is fairly poor. There are hundreds of billions of dollars of capital flowing around, but public goods that are key to that capital's ongoing survival are receiving only tens of millions of dollars per year of funding.

There are two ways to respond to this fact. The first way is to be proud of these limitations and the valiant, even if not particularly effective, efforts that your community makes to work around them. This seems to be the route that the Bitcoin ecosystem often takes:

Collaborating for Philanthropy

This is why **zkSNACKs**, alongside **Francis Pouliot, CEO of Bull Bitcoin**, have come together to make a .86 bitcoin, or $40,000 contribution (split evenly between the two companies) in **support of the growth and development of Bitcoin Knots** - an open source enhanced bitcoin node/wallet software. More specifically, Bitcoin Knots is a Bitcoin full node and wallet software which can be used as an alternative to the more popular Bitcoin Core.

One of Bull Bitcoin's core values is "skin in the game".

> *Cypherpunks write code, but cypherpunks don't always get paid. We can't expect the world's most talented experts to contribute indefinitely without financial compensation. If the companies that profit from Bitcoin open-source development don't provide the necessary funding, who will? ~ Francis Pouliot*

The personal self-sacrifice of the teams funding core development is of course admirable, but it's admirable the same way that Eliud Kipchoge running a marathon in under two hours is admirable: it's an impressive show of human fortitude, but it's not the future of transportation (or, in this case, public-goods funding). Much like we have better technologies to allow people to move forty-two kilometers in under an hour without exceptional fortitude and years of training, **we should also focus on building better social technologies to fund public goods at the scales that we need, and as a systemic part of our economic ecology and not one-off acts of philanthropic initiative.**

Now, let us get back to cryptocurrency. A major power of crypto-currency (and other digital assets such as domain names, virtual land, and NFTs) is that it allows communities to summon up large amounts of capital without any individual person needing to personally donate that capital. However, *this capital is constrained by conceptions of legitimacy:* you cannot simply allocate it to a centralized team without compromising on what makes it valuable. While Bitcoin and Ethereum do already rely on conceptions of legitimacy to respond to 51% attacks, using conceptions of legitimacy to guide in-protocol funding of public goods is much harder. But at the increasingly rich application layer where new protocols are constantly being created, we have quite a bit more flexibility in where that funding could go.

LEGITIMACY IN BITSHARES

One of the long-forgotten, but in my opinion very innovative, ideas from the early cryptocurrency space was the BitShares social-consensus model. Essentially, BitShares described itself as a community of people (PTS and AGS holders) who were willing to help collectively support an ecosystem of new projects, but for a project to be welcomed into the ecosystem, it would have to allocate 10% of its token supply to existing PTS and AGS holders.

Now, of course anyone can make a project that does not allocate any coins to PTS/AGS holders, or even fork a project that did make an allocation and take the allocation out. But as Dan Larimer says:

> You cannot force anyone to do anything, but in this market it is all network effect. If someone comes up with a compelling implementation then you can adopt the entire PTS community for the cost of generating a new genesis block. The individual who decided to start from scratch would have to build an entire new community

around his system. Considering the network effect, I sus-
pect that the coin that honors ProtoShares will win.

This is also a conception of legitimacy: any project that makes the
allocation to PTS/AGS holders will get the attention and support
of the community (and it will be worthwhile for each individual
community member to take an interest in the project because the
rest of the community is doing so as well), and any project that
does not make the allocation will not. **Now, this is certainly not a
conception of legitimacy that we want to replicate verbatim—
there is little appetite in the Ethereum community for enriching
a small group of early adopters—but the core concept can be
adapted into something much more socially valuable.**

EXTENDING THE MODEL TO ETHEREUM

Blockchain ecosystems, Ethereum included, value freedom and
decentralization. But the public-goods ecology of most of these
blockchains is, regrettably, still quite authority-driven and centralized: whether it's Ethereum, Zcash, or any other major blockchain,
there is typically one entity (or at most two to three) that far out-
spends everyone else, giving independent teams that want to build
public goods few options. I call this model of public-goods funding
"Central Capital Coordinators for Public-goods" (CCCPs).

This state of affairs is not the fault of the organizations themselves, since they are typically valiantly doing their best to support the ecosystem. Rather, it's the rules of the ecosystem that are being *unfair to that organization*, because they hold the organization to an unfairly high standard. Any single centralized organization will inevitably have blind spots and at least a few categories and teams whose value it fails to understand; this is not because anyone involved is doing anything wrong, but because such perfection is beyond the reach of small groups of humans. So there is great value in creating a more diversified and resilient approach to public-goods funding to take the pressure off any single organization.

Fortunately, we already have the seed of such an alternative! The Ethereum application-layer ecosystem exists, is growing increasingly powerful, and is already showing its public-spiritedness. Companies like Gnosis have been contributing to Ethereum client development, and various Ethereum DeFi* projects have donated hundreds of thousands of dollars to the Gitcoin Grants matching pool.

Gitcoin Grants has already achieved a high level of legitimacy: its public-goods-funding mechanism, quadratic funding, has proven itself to be credibly neutral and effective at reflecting the community's priorities and values and plugging the holes left by

---

* DeFi refers to "decentralized finance": financial instruments and applications that operate on blockchain networks.

existing funding mechanisms. Sometimes, top Gitcoin Grants matching recipients are even used as inspiration for grants by other and more centralized grant-giving entities. The Ethereum Foundation itself has played a key role in supporting this experimentation and diversity, incubating efforts like Gitcoin Grants, along with MolochDAO and others, that then go on to get broader community support.

We can make this nascent public-goods-funding ecosystem even stronger by taking the BitShares model, and making a modification: instead of giving the strongest community support to projects that allocate tokens to a small oligarchy who bought PTS or AGS back in 2013, **we support projects that contribute a small portion of their treasuries toward the public goods that make them, and the ecosystem that they depend on, possible**. And, crucially, we can deny these benefits to projects that fork an existing project and do not give back value to the broader ecosystem.

There are many ways to support public goods: making a long-term commitment to support the Gitcoin Grants matching pool, supporting Ethereum client development (also a reasonably credibly-neutral task, as there's a clear definition of what an Ethereum client is), or even running one's own grant program whose scope goes beyond that particular application-layer project itself. The easiest way to agree on what counts as sufficient support is to agree on how much—for example, 5% of a project's spending going to support the broader ecosystem and another 1% going to public goods that go beyond the blockchain space—and rely on good faith to choose where that funding would go.

DOES THE COMMUNITY ACTUALLY HAVE THAT MUCH LEVERAGE?

Of course, there are limits to the value of this kind of community support. If a competing project (or even a fork of an existing project) gives its users a much better offering, then users are going to flock to it, regardless of how many people yell at them to instead use some alternative that they consider to be more pro-social.

But these limits are different in different contexts; sometimes the community's leverage is weak, but at other times it's quite strong. An interesting case study in this regard is the case of Tether vs. DAI. Tether has many scandals, but despite this, traders use Tether to hold and move around dollars all the time. The more decentralized and transparent DAI, despite its benefits, is unable to take away much of Tether's market share, at least as far as traders go. But where DAI excels is applications: Augur uses DAI, xDai uses DAI, PoolTogether uses DAI, zk.money plans to use DAI, and the list goes on. What dapps use USDT? Far fewer.

Hence, though the power of community-driven legitimacy effects is not infinite, there is nevertheless considerable room for leverage, enough to encourage projects to direct at least a small percent of their budgets to the broader ecosystem. There's even a selfish reason to participate in this equilibrium: if you were the developer of an Ethereum wallet, or an author of a podcast or newsletter, and you saw two competing projects, one of which contributes significantly to ecosystem-level public goods and one of which does not, which one would you do your utmost to help secure more market share?

NFTS: SUPPORTING PUBLIC GOODS BEYOND ETHEREUM

The concept of supporting public goods through value generated "out of the ether" by publicly supported conceptions of legitimacy has value going far beyond the Ethereum ecosystem. An important and immediate challenge and opportunity is NFTs. NFTs stand a great chance of significantly helping many kinds of public goods, especially of the creative variety, at least partially solve their chronic and systemic funding deficiencies.

Jack Dorsey's first tweet may fetch $2.5 million, and he'll donate the NFTy proceeds to charity

The auction ends on March 21st

By Jay Peters | @jaypeters | Mar 9, 2021, 12:06pm EST

f 🐦 ⌐ SHARE

Actually, a very admirable first step.

But there could also be a missed opportunity: there is little social value in helping Elon Musk earn yet another $1 million by selling his tweet when, as far as we can tell, the money is just going to himself (and, to his credit, he eventually decided not to sell). If NFTs simply become a casino that largely benefits already-wealthy celebrities, that would be a far less interesting outcome.

Fortunately, we have the ability to help shape the outcome. Which NFTs people find attractive to buy, and which ones they do not, is a question of legitimacy: if everyone agrees that one NFT is interesting and another NFT is not, then people will strongly prefer buying the first, because it would have both higher value for bragging rights and personal pride in holding it, and because it could be resold for more since everyone else is thinking in the same way. If the conception of legitimacy for NFTs can be pulled in a good direction, there is an opportunity to establish a solid channel of funding to artists, charities, and others.

Here are two potential ideas:

1. Some institution (or even DAO) could "bless" NFTs in exchange for a guarantee that some portion of the revenues goes toward a charitable cause, ensuring that multiple groups benefit at the same time. This blessing could even come with an official categorization: Is the NFT dedicated to global poverty relief, scientific research, creative arts, local journalism, open-source software development, empowering marginalized communities, or something else?

2. We can work with social media platforms to make NFTs more visible on people's profiles, giving buyers a way to show the values that they committed not just their words but their hard-earned money to. This could be combined with (1) nudging users toward NFTs that contribute to valuable social causes.

There are definitely more ideas, but this is an area that certainly deserves more active coordination and thought.

IN SUMMARY

□ **Legitimacy (higher-order acceptance) is very powerful.** Legitimacy appears in any context where there is coordination, and especially on the internet, coordination is everywhere.

□ There are different ways in which legitimacy comes to be: **brute force, continuity, fairness, process, performance, and participation** are among the important ones.

□ Cryptocurrency is powerful because it lets us summon up large pools of capital by collective economic will, and these pools of capital are, at the beginning, not controlled by any person. Rather, these **pools of capital are *controlled directly by concepts of legitimacy*.**

☐ It's too risky to start doing public-goods funding by printing tokens at the base layer. Fortunately, however, Ethereum has a very rich **application-layer ecosystem**, where we have much more flexibility. This is in part because there's an opportunity not just to influence existing projects, but also shape new ones that will come into existence in the future.

☐ **Application-layer projects that support public goods in the community should get the support of the community**, and this is a big deal. The example of DAI* shows that this support really matters!

☐ The Ethereum ecosystem cares about mechanism design and innovating at the social layer. The Ethereum ecosystem's own public-goods funding challenges are a great place to start!

☐ But this goes far beyond just Ethereum itself. NFTs are one example of a large pool of capital that depends on concepts of legitimacy. **The NFT industry could be a significant boon** to artists, charities, and other public-goods providers far beyond our own virtual corner of the world, but **this outcome is not predetermined; it depends on active coordination and support**.

Special thanks to Karl Floersch, Aya Miyaguchi, and Mr. Silly for ideas, feedback, and review.

---

* As mentioned above in passing, DAI's parent MolochDAO received early funding from the public-goods-focused Gitcoin Grants program.

AGAINST OVERUSE OF THE GINI COEFFICIENT

---

vitalik.ca
July 29, 2021

The Gini coefficient (also called the Gini index) is by far the most popular and widely known measure of inequality, typically used to measure inequality of income or wealth in some country, territory, or other community. It's popular because it's easy to understand, with a mathematical definition that can easily be visualized on a graph.

However, as one might expect from any scheme that tries to reduce inequality to a single number, the Gini coefficient also has its limits. This is true even in its original context of measuring income and wealth inequality in countries, but it becomes even more true when the Gini coefficient is transplanted into other contexts (particularly: cryptocurrency). In this post I will talk about some of the limits of the Gini coefficient, and propose some alternatives.

WHAT IS THE GINI COEFFICIENT?

The Gini coefficient is a measure of inequality introduced by Corrado Gini in 1912. It is typically used to measure inequality of income and wealth of countries, though it is also increasingly being used in other contexts.

There are two equivalent definitions of the Gini coefficient:

☐ **AREA-ABOVE-CURVE DEFINITION:** Draw the graph of a function, where $f(p)$ equals the share of total income earned by the lowest-earning portion of the population (e.g., $f(0.1)$ is the share of total income earned by the lowest-earning 10%). The Gini coefficient is the area between that curve and the $y = x$ line, as a portion of the whole triangle:

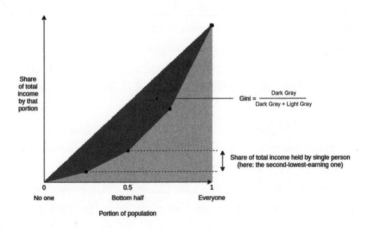

☐ **AVERAGE-DIFFERENCE DEFINITION:** The Gini coefficient is half the average difference of incomes between all possible pairs of individuals, divided by the mean income.

For example, in the above example chart, the four incomes are [1, 2, 4, 8], so the sixteen possible differences are [0, 1, 3, 7, 1, 0, 2, 6, 3, 2, 0, 4, 7, 6, 4, 0]. Hence the average difference is 2.875 and the mean income is 3.75, so Gini =

$$\frac{2.875}{2 \times 3.75} \approx 0.3833$$

It turns out that the two are mathematically equivalent (proving this is an exercise to the reader)!

WHAT'S WRONG WITH THE GINI COEFFICIENT?

The Gini coefficient is attractive because it's a reasonably simple and easy-to-understand statistic. It might not look simple, but trust me, pretty much everything in statistics that deals with populations of arbitrary size is that bad, and often much worse. Here, stare at the formula of something as basic as the standard deviation:

$$\sigma = \frac{\sum_{i=1}^{n} x_i^2}{n} - (\frac{\sum_{i=1}^{n} x_i}{n})^2$$

And here's the Gini:

$$G = \frac{2 \times \sum_{i=1}^{n} i \times x_i}{n \times \sum_{i=1}^{n} x_i} - \frac{n+1}{n}$$

It's actually quite tame, I promise!

So, what's wrong with it? Well, there are lots of things wrong with it, and people have written lots of articles about various problems with the Gini coefficient. In this article, I will focus on one specific problem that I think is under-discussed about the

Gini as a whole, but that has particular relevance to analyzing inequality in internet communities. **The Gini coefficient combines together into a single inequality index two problems that actually look quite different: suffering due to lack of resources and concentration of power.**

To understand the difference between the two problems more clearly, let's look at two dystopias:

□ **DYSTOPIA A:** Half the population equally shares all the resources; everyone else has none.

□ **DYSTOPIA B:** One person has half of all the resources; everyone else equally shares the remaining half.

Here are the Lorenz curves (fancy charts like we saw above) for both dystopias:

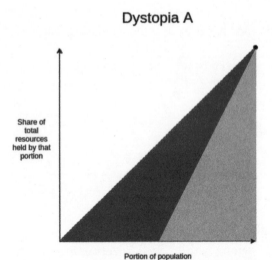

Dystopia A

Share of total resources held by that portion

Portion of population

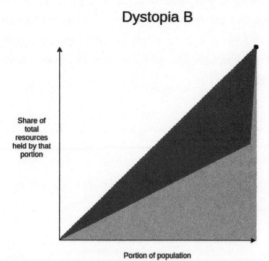

Dystopia B

Share of total resources held by that portion

Portion of population

Clearly, neither of those two dystopias is a good place to live. But they are not-very-nice places to live in very different ways. Dystopia A gives each resident a coin flip between unthinkably horrific mass starvation if they end up on the left half of the distribution and egalitarian harmony if they end up on the right half. If you're Thanos,* you might actually like it! If you're not, it's worth avoiding with the strongest force. Dystopia B, on the other hand, is *Brave New World*–like: everyone has decently good lives (at least at the time when that snapshot of everyone's resources is taken), but at the high cost of an extremely undemocratic power structure where you'd better hope you have a good overlord. If you're Curtis Yarvin,** you might actually like it! If you're not, it's very much worth avoiding too.

These two problems are different enough that they're worth

---

* A Marvel Comics character who killed half the population of the universe in order to impress Mistress Death.
** A neo-monarchist blogger who developed Urbit, a peer-to-peer server platform.

analyzing and measuring separately. And this difference is not just theoretical. Here is a chart showing share of total income earned by the bottom 20% (a decent proxy for avoiding dystopia A) versus share of total income earned by the top 1% (a decent proxy for being near dystopia B):

Sources: https://data.worldbank.org/indicator/SI.DST.FRST.20 (merging 2015 and 2016 data) and http://hdr.undp.org/en/indicators/186106.

The two are clearly correlated (coefficient -0.62), but very far from perfectly correlated (the high priests of statistics apparently consider 0.7 to be the lower threshold for being "highly correlated," and we're even under that). There's an interesting second dimension to the chart that can be analyzed—what's the difference between a country where the top 1% earns 20% of the total income and the bottom 20% earns 3% and a country where the top 1% earns 20% and the bottom 20% earns 7%? Alas, such an exploration is best left to other enterprising data and culture explorers with more experience than myself.

WHY GINI IS VERY PROBLEMATIC IN NON-GEOGRAPHIC COMMUNITIES (E.G., INTERNET/CRYPTO COMMUNITIES)

Wealth concentration within the blockchain space in particular is an important problem, and it's a problem worth measuring and understanding. It's important for the blockchain space as a whole, as many people (and US Senate hearings) are trying to figure out to what extent crypto is truly anti-elitist and to what extent it's just replacing old elites with new ones. It's also important when comparing different cryptocurrencies with each other.

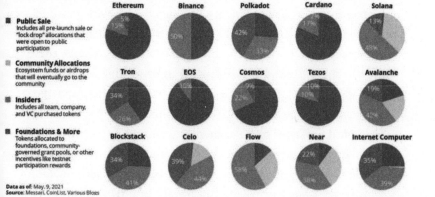

Share of coins explicitly allocated to specific insiders in a cryptocurrency's initial supply is one type of inequality. Note that the Ethereum data is slightly wrong: the insider and foundation shares should be 12.3% and 4.2%, not 15% and 5%.

Given the level of concern about these issues, it should be not at all surprising that many people have tried computing Gini indices of cryptocurrencies. They are not all as bad as when we had to deal with this sensationalist article from 2014:

How Bitcoin Is Like North Korea

Joe Weisenthal Jan 13, 2014, 12:04 AM

Citigroup currency analyst Steven Englander is out with a long Sunday note talking about everyone's favorite topic: digital currency.

In it, he makes an important observation about the extreme inequality in the Bitcoin world:

North Korea's Korean Central News Agency/AP

In addition to common plain methodological mistakes (mixing up income with wealth inequality, mixing up users with accounts, or both) that such analyses make quite frequently, there is a deep and subtle problem with using the Gini coefficient to make these kinds of comparisons. The problem lies in key distinction between typical geographic communities (e.g., cities, countries) and typical internet communities (e.g., blockchains):

A typical resident of a geographic community spends most of their time and resources in that community, and so measured inequality in a geographic community reflects inequality in total resources available to people. **But in an internet community, measured inequality can come from two sources: (i) inequality in total resources available to different participants, and (ii) inequality in level of interest in participating in the community.**

The average person with \$15 in fiat currency is poor and is missing out on the ability to have a good life. The average person with \$15 in cryptocurrency is a dabbler who opened up a wallet once for fun. Inequality in level of interest is a healthy thing; every community has its dabblers and its full-time hardcore fans with no life. So if a cryptocurrency has a very high Gini coefficient, but it turns out that much of this inequality comes from inequality in level of interest, then the number points to a much less scary reality than the headlines imply.

Cryptocurrencies, even those that turn out to be highly plutocratic, will not turn any part of the world into anything close to dystopia A. But badly distributed cryptocurrencies may well look like dystopia B, a problem compounded if coin-voting governance is used to make protocol decisions. Hence, to detect the problems that cryptocurrency communities worry about most, we want a metric that captures proximity to dystopia B more specifically.

AN ALTERNATIVE: MEASURING DYSTOPIA A PROBLEMS AND DYSTOPIA B PROBLEMS SEPARATELY

An alternative approach to measuring inequality involves directly estimating suffering from resources being unequally distributed (that is, "dystopia A" problems). First, start with some utility function representing the value of having a certain amount of money; $\log(x)$ is popular, because it captures the intuitively appealing approximation that doubling one's income is about as useful at any level: going from \$10,000 to \$20,000 adds the same utility as going from \$5,000 to \$10,000 or from \$40,000 to \$80,000. The score is then a matter of measuring how much utility is lost compared to if everyone just got the average income:

$$\log\left(\frac{\sum_{i=1}^{n} x_i}{n}\right) - \frac{\sum_{i=1}^{n} \log(x_i)}{n}$$

The first term (log-of-average) is the utility that everyone would have if money were perfectly redistributed, so everyone earned the average income. The second term (average-of-log) is the average utility in that economy today. The difference represents lost utility from inequality, if you look narrowly at resources as something used for personal consumption. There are other ways to define this formula, but they end up being close to equivalent (e.g., the 1969 paper by Anthony Atkinson suggested an "equally distributed equivalent level of income" metric, which, in the $U(x) = \log(x)$ case, is just a monotonic function of the above, and the Theil L index is perfectly mathematically equivalent to the above formula).

To measure concentration (or "dystopia B" problems), the Herfindahl-Hirschman index is an excellent place to start, and is already used to measure economic concentration in industries:

$$\frac{\sum_{i=1}^{n} x_i^2}{\left(\sum_{i=1}^{n} x_i\right)^2}$$

Or for you visual learners out there:

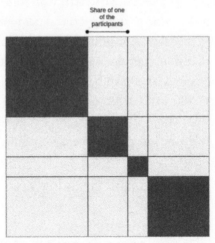

Herfindahl-Hirschman index: dark gray area
divided by total area.

There are other alternatives to this; the Theil T index has some similar properties though also some differences. A simpler and dumber alternative is the Nakamoto coefficient: the minimum number of participants needed to add up to more than 50% of the total. Note that all three of these concentration indices focus heavily on what happens near the top (and deliberately so): a large number of dabblers with a small quantity of resources contributes little or nothing to the index, while the act of two top participants merging can make a very big change to the index.

For cryptocurrency communities—where concentration of resources is one of the biggest risks to the system, but someone having only 0.00013 coins is not any kind of evidence that they're actually starving—adopting indices like this is the obvious approach. But even for countries, it's probably worth talking about, and measuring, concentration of power and suffering from lack of resources more separately.

That said, **at some point we have to move beyond even these indices**. The harms from concentration are not just a function of the size of the actors; they are also heavily dependent on the relationships between the actors and their ability to collude with each other. Similarly, resource allocation is network-dependent: lack of formal resources may not be that harmful if the person lacking resources has an informal network to tap into. But dealing with these issues is a much harder challenge, and so we do also need the simpler tools while we still have less data to work with.

Special thanks to Barnabé Monnot and Tina Zhen for feedback and review.

MOVING BEYOND COIN-VOTING GOVERNANCE

vitalik.ca
August 16, 2021

One of the important trends in the blockchain space over the past year is the transition from focusing on **decentralized finance (DeFi)** to also thinking about **decentralized governance (DeGov)**. While 2020 is often widely, and with much justification, hailed as a year of DeFi, over the years since then the growing complexity and capability of DeFi projects that make up this trend has led to growing interest in decentralized governance to handle that complexity. There are examples inside of Ethereum: YFI, Compound, Synthetix, UNI, Gitcoin and others have all launched, or even started with, some kind of DAO. But it's also true outside of Ethereum, with arguments over infrastructure funding proposals in Bitcoin Cash, infrastructure funding votes in Zcash, and much more.

The rising popularity of formalized decentralized governance is undeniable, and there are important reasons why people are interested in it. But it is also important to keep in mind the risks of such schemes, as the recent hostile takeover of Steem and subsequent mass exodus to Hive make clear. I would further argue that these trends are unavoidable. **Decentralized governance in some contexts is both necessary and dangerous**, for reasons that I will get into in this post. How can we get the benefits of DeGov while minimizing the risks? I will argue for one key part of the answer: **we need to move beyond coin voting as it exists in its present form**.

DEGOV IS NECESSARY

Ever since the Declaration of Independence of Cyberspace in 1996,* there has been a key unresolved contradiction in what can be called cypherpunk ideology. On the one hand, cypherpunk values are all about using cryptography to minimize coercion, and maximize the efficiency and reach of the main non-coercive coordination mechanism available at the time: private property and markets. On the other hand, the economic logic of private property and markets is optimized for activities that can be "decomposed" into repeated one-to-one interactions, and the infosphere, where art, documentation, science, and code are produced and consumed through irreducibly one-to-many interactions, is the exact opposite of that.

There are two key problems inherent to such an environment that need to be solved:

☐ **FUNDING PUBLIC GOODS:** How do projects that are valuable to a wide and unselective group of people in the community, but which often do not have a business model (e.g., layer 1 and layer 2 protocol research, client development, documentation . . .), get funded?

---

* This was a statement issued from the World Economic Forum in Davos by John Perry Barlow, an early internet advocate and former Grateful Dead lyricist, on the occasion of the passage of restrictive regulations by the US Congress.

□   **PROTOCOL MAINTENANCE AND UPGRADES:** How are upgrades
to the protocol, and regular maintenance and adjustment
operations on parts of the protocol that are not long-term
stable (e.g., lists of safe assets, price oracle sources, multiparty computation keyholders), agreed upon?

Early blockchain projects largely ignored both of these challenges, pretending that the only public good that mattered was
network security, which could be achieved with a single algorithm set in stone forever and paid for with fixed proof-of-work
rewards. This state of affairs in funding was possible at first
because of extreme Bitcoin price rises from 2010–13, then the
one-time ICO boom from 2014–17, and again from the simultaneous second crypto bubble of 2014–17, all of which made the
ecosystem wealthy enough to temporarily paper over the large
market inefficiencies. Long-term governance of public resources
was similarly ignored: Bitcoin took the path of extreme minimization, focusing on providing a fixed-supply currency and ensuring
support for layer 2 payment systems like Lightning and nothing
else. Ethereum continued developing mostly harmoniously (with
one major exception)* because of the strong legitimacy of its pre-
existing road map (basically: "proof of stake and sharding"), and
sophisticated application-layer projects that required anything
more did not yet exist.

But now, increasingly, that luck is running out, and the challenges of coordinating protocol maintenance and upgrades and
funding documentation and research and development, while
avoiding the risks of centralization, are at the forefront.

---

* That is, the DAO hack.

THE NEED FOR DEGOV FOR FUNDING PUBLIC GOODS

It is worth stepping back and seeing the absurdity of the present situation. Daily mining issuance rewards from Ethereum are about 13,500 ETH, or about $40 million, per day. Transaction fees are similarly high; the non-EIP-1559-burned portion* continues to be around 1,500 ETH (about $4.5 million) per day. So there are many billions of dollars per year going to fund network security. Now, what is the budget of the Ethereum Foundation? About $30–60 million per year. There are non-EF actors (e.g., ConsenSys) contributing to development, but they are not much larger. The situation in Bitcoin is similar, with perhaps even less funding going into non-security public goods.

Here is the situation in a familiar chart:

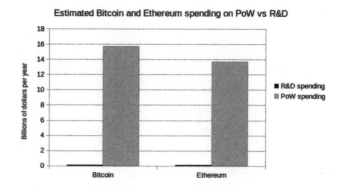

Within the Ethereum ecosystem, one can make a case that this disparity does not matter too much; tens of millions of dollars per year is "enough" to do the needed R&D and adding more funds does not necessarily improve things, and so the risks to the plat-

---

* This refers to a 2021 "Ethereum Improvement Proposal" that changed the structure of the gas-fee market.

form's credible neutrality from instituting in-protocol developer funding exceed the benefits. But in many smaller ecosystems, both ecosystems within Ethereum and those of entirely separate blockchains like BCH and Zcash, the same debate is brewing, and at those smaller scales the imbalance makes a big difference.

Enter DAOs. A project that launches as a "pure" DAO from day one can achieve a combination of two properties that were previously impossible to combine: (i) sufficiency of developer funding, and (ii) credible neutrality of funding (the much-coveted "fair launch"). Instead of developer funding coming from a hardcoded list of receiving addresses, the decisions can be made by the DAO itself.

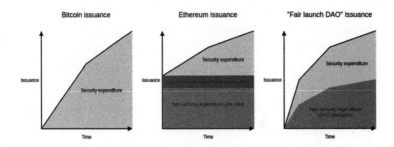

Of course, it's difficult to make a launch perfectly fair, and unfairness from information asymmetry can often be worse than unfairness from explicit premines (was Bitcoin really a fair launch considering how few people had a chance to even hear about it by the time one-fourth of the supply had already been handed out by the end of 2010?). But even still, in-protocol compensation for non-security public goods from day one seems like a potentially significant step forward toward getting sufficient and more credibly neutral developer funding.

THE NEED FOR DEGOV FOR PROTOCOL MAINTENANCE AND UPGRADES

In addition to public-goods funding, the other equally important problem requiring governance is protocol maintenance and upgrades. While I advocate trying to minimize all nonautomated parameter adjustment (see the "limited governance" section below) and I am a fan of RAI's "un-governance" strategy, there are times where governance is unavoidable. Price oracle inputs must come from somewhere, and occasionally that somewhere needs to change. Until a protocol "ossifies" into its final form, improvements have to be coordinated somehow. Sometimes, a protocol's community might *think* that they are ready to ossify, but then the world throws a curveball that requires a complete and controversial restructuring. What happens if the US dollar collapses, and RAI has to scramble to create and maintain their own decentralized CPI index* for their stablecoin to remain stable and relevant? Here, too, DeGov is necessary, and so avoiding it outright is not a viable solution.

One important distinction is whether or not *off-chain* governance** is possible. I have for a long time been a fan of off-chain governance wherever possible. And indeed, for base-layer blockchains, off-chain governance absolutely is possible. **But for application-layer projects, and especially DeFi projects, we run into the problem that application-layer smart-contract systems *often directly control external assets*, and that control cannot be forked away.** If Tezos's on-chain governance gets cap-

---

* CPI stands for consumer price index. RAI is stablecoin but (unlike DAI and USDT) is not pegged to a "fiat" currency like the US dollar. It seeks greater stability while still being reflective of changes in the underlying crypto markets.

** Whereas "on-chain" governance refers to voting and other decision-making through blockchain protocols directly, "off-chain" can refer to mechanisms like foundations and companies, oligarchic control over a DAO, informal charismatic authority, whisper networks, and more.

tured by an attacker, the community can hard fork away without any losses beyond (admittedly high) coordination costs. If MakerDAO's on-chain governance gets captured by an attacker, the community can absolutely spin up a new MakerDAO, but they will lose all the ETH and other assets that are stuck in the existing MakerDAO CDPs. **Hence, while off-chain governance is a good solution for base layers and some application-layer projects, many application-layer projects, particularly DeFi, will inevitably require formalized on-chain governance of some form.**

DEGOV IS DANGEROUS

However, all current instantiations of decentralized governance come with great risks. To followers of my writing, this discussion will not be new. There are two primary types of issues with coin voting that I worry about: (i) inequalities and incentive misalignments even in the absence of attackers, and (ii) outright attacks through various forms of (often obfuscated) vote buying. To the former, there have already been many proposed mitigations (e.g., delegation), and there will be more. But the latter is a much more dangerous elephant in the room to which I see no solution within the current coin-voting paradigm.

PROBLEMS WITH COIN VOTING EVEN IN THE ABSENCE OF ATTACKERS

The problems with coin voting even without explicit attackers are increasingly well-understood, and mostly fall into a few buckets:

□ **Small groups of wealthy participants ("whales") are better at successfully executing decisions than large groups of small-holders**: This is because of the tragedy

of the commons among small-holders: each small-holder
has only an insignificant influence on the outcome, and
so they have little incentive to not be lazy and actually
vote. Even if there are rewards for voting, there is little
incentive to research and think carefully about what they
are voting for.

☐ **Coin-voting governance empowers coin holders and
coin holder interests at the expense of other parts of the
community**: Protocol communities are made up of diverse
constituencies that have many different values, visions
and goals. Coin voting, however, only gives power to one
constituency (coin holders, and especially wealthy ones),
and leads to overvaluing the goal of making the coin price
go up even if that involves harmful rent extraction.

☐ **Conflict of interest issues**: Giving voting power to one
constituency (coin holders), and especially over-empowering
wealthy actors in that constituency, risks overexposure to
the conflicts of interest within that particular elite (e.g.,
investment funds or holders that also hold tokens of
other DeFi platforms that interact with the platform in
question).

There is one major type of strategy being attempted for
solving the first problem (and therefore also mitigating the third
problem): delegation. Small-holders don't have to personally
judge each decision; instead, they can delegate to community
members that they trust. This is an honorable and worthy
experiment; we shall see how well delegation can mitigate the
problem.

My voting-delegation page in the Gitcoin DAO.

The problem of coin-holder centrism, on the other hand, is significantly more challenging: coin-holder centrism is inherently baked into a system where coin-holder votes are the only input. The misperception that coin-holder centrism is an intended goal, and not a bug, is already causing confusion and harm; one (broadly excellent) article* discussing blockchain public goods complains:

> Can crypto protocols be considered public goods if ownership is concentrated in the hands of a few whales? Colloquially, these market primitives are sometimes described as "public infrastructure," but if blockchains serve a "public" today, it is primarily one of decentralized finance. Fundamentally, these tokenholders share only one common object of concern: price.

---

* Sam Hart, Laura Lotti, and Toby Shorin, "Positive Sum Worlds: Remaking Public Goods," Other Internet, July 2, 2021.

The complaint is false; blockchains serve a public much richer and broader than DeFi token holders. But our coin-voting-driven governance systems are completely failing to capture that, and it seems difficult to make a governance system that captures that richness without a more fundamental change to the paradigm.

COIN VOTING'S DEEP, FUNDAMENTAL VULNERABILITY TO ATTACKERS: VOTE BUYING

The problems get much worse once determined attackers trying to subvert the system enter the picture. The fundamental vulnerability of coin voting is simple to understand. **A token in a protocol with coin voting is a bundle of two rights that are combined into a single asset: (i) some kind of economic interest in the protocol's revenue and (ii) the right to participate in governance. This combination is deliberate: the goal is to align power and responsibility. But in fact, these two rights are very easy to unbundle from each other.** Imagine a simple wrapper contract that has these rules: if you deposit 1 XYZ into the contract, you get back 1 WXYZ. That WXYZ can be converted back into an XYZ at any time, plus it accrues dividends. Where do the dividends come from? Well, while the XYZ coins are inside the wrapper contract, it's the wrapper contract that has the ability to use them however it wants in governance (making proposals, voting on proposals, etc.). The wrapper contract simply auctions off this right every day, and distributes the profits among the original depositors.

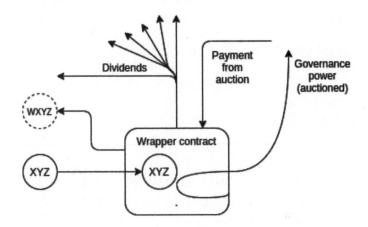

As an XYZ holder, is it in your interest to deposit your coins into the contract? If you are a very large holder, it might not be; you like the dividends, but you are scared of what a misaligned actor might do with the governance power you are selling them. But if you are a smaller holder, then it very much is. If the governance power auctioned by the wrapper contract gets bought up by an attacker, you personally only suffer a small fraction of the cost of the bad governance decisions that your token is contributing to, but you personally gain the full benefit of the dividend from the governance rights auction. This situation is a classic tragedy of the commons.

Suppose that an attacker makes a decision that corrupts the DAO to the attacker's benefit. The harm per participant from the decision succeeding is D, and the chance that a single vote tilts the outcome is p. Suppose an attacker makes a bribe of B. The game chart looks like this:

Decision	Benefit to you	Benefit to others
Accept attacker's bribe	$B - D \times p$	$-999 \times D \times p$
Reject bribe, vote your conscience	0	0

If $B > (D \times p)$, you are inclined to accept the bribe, but as long as $B < (1000 \times D \times p)$, accepting the bribe is *collectively* harmful. So if $p < 1$ (usually, p is far below 1), there is an opportunity for an attacker to bribe users to adopt a net-negative decision, compensating each user far less than the harm they suffer.

One natural critique of voter bribing fears is: Are voters *really* going to be so immoral as to accept such obvious bribes? The average DAO token holder is an enthusiast, and it would be hard for them to feel good about so selfishly and blatantly selling out the project. But what this misses is that there are much more obfuscated ways to separate out profit-sharing rights from governance rights, which don't require anything remotely as explicit as a wrapper contract.

The simplest example is borrowing from a DeFi lending platform (e.g., Compound). Someone who already holds ETH can lock up their ETH in a CDP ("collateralized debt position") in one of these platforms, and once they do that the CDP contract allows them to borrow an amount of XYZ up to, for example, half the value of the ETH that they put in. They can then do whatever they want with this XYZ. To recover their ETH, they would eventually need to pay back the XYZ that they borrowed, plus interest.

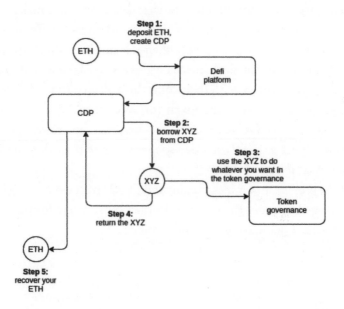

Note that throughout this process, *the borrower has no financial exposure to XYZ*. That is, if they use their XYZ to vote for a governance decision that destroys the value of XYZ, they do not lose a penny as a result. The XYZ they are holding is XYZ that they have to eventually pay back into the CDP regardless, so they do not care if its value goes up or down. **And so we have achieved unbundling: the borrower has governance power without economic interest, and the lender has economic interest without governance power.**

There are also centralized mechanisms for separating profit-sharing rights from governance rights. Most notably, when users deposit their coins on a (centralized) exchange, the exchange holds full custody of those coins, and the exchange has the ability to use those coins to vote. This is not mere theory; there is evidence of exchanges using their users' coins in several DPoS systems.

Some DAO protocols are using timelock techniques to limit these attacks, requiring users to lock their coins and make them immovable for some period of time in order to vote. These techniques can limit buy-then-vote-then-sell attacks in the short term, but ultimately timelock mechanisms can be bypassed by users holding and voting with their coins through a contract that issues a wrapped version of the token (or, more trivially, a centralized exchange). **As far as security mechanisms go, timelocks are more like a paywall on a newspaper website than they are like a lock and key.**

At present, many blockchains and DAOs with coin voting have managed to avoid these attacks in their most severe forms. There are occasional signs of attempted bribes:

But despite all of these important issues, there have been much fewer examples of outright voter bribing, including obfuscated forms such as using financial markets, than simple economic reasoning would suggest. The natural question to ask is: Why haven't more outright attacks happened yet?

My answer is that the "why not yet" relies on three contingent factors that are true today, but are likely to get less true over time:

1. **COMMUNITY SPIRIT:** Having a tightly knit community, where everyone feels a sense of camaraderie in a common tribe and mission.

2. **HIGH WEALTH CONCENTRATION AND COORDINATION OF TOKEN HOLDERS:** Large holders have higher ability to affect the outcome and have investments in long-term relationships with each other (both the "old boys clubs" of VCs, but also many other equally powerful but lower-profile groups of wealthy token holders), and this makes them much more difficult to bribe.

3. **IMMATURE FINANCIAL MARKETS IN GOVERNANCE TOKENS:** Ready-made tools for making wrapper tokens exist in proof-of-concept forms but are not widely used, bribing contracts exist but are similarly immature, and liquidity in lending markets is low.

When a small coordinated group of users holds over 50% of the coins, *and* both they and the rest are invested in a tightly knit community, and there are few tokens being lent out at reasonable rates, all of the above bribing attacks may perhaps remain theoretical. But over time, (1) and (3) will inevitably become less true no matter what we do, and (2) *must* become less true if we want DAOs to become more fair. When those changes happen, will

DAOs remain safe? And if coin voting cannot be sustainably resistant against attacks, then what can?

SOLUTION 1: LIMITED GOVERNANCE

One possible mitigation to the above issues, and one that is to varying extents being tried already, is to put limits on what coin-driven governance can do. There are a few ways to do this:

☐ **USE ON-CHAIN GOVERNANCE ONLY FOR APPLICATIONS, NOT BASE LAYERS:** Ethereum does this already, as the protocol itself is governed through off-chain governance, while DAOs and other apps on top of this are sometimes (but not always) governed through on-chain governance.

☐ **LIMIT GOVERNANCE TO FIXED PARAMETER CHOICES:** Uniswap does this, as it only allows governance to affect (i) token distribution and (ii) a 0.05% fee in the Uniswap exchange. Another great example is RAI's "un-governance" road map, where governance has control over fewer and fewer features over time.

☐ **ADD TIME DELAYS:** A governance decision made at time T takes effect only at, for example, T + 90 days. This allows users and applications that consider the decision unacceptable to move to another application (possibly a fork). Compound has a time-delay mechanism in its governance, but in principle the delay can (and eventually should) be much longer.

☐ **BE MORE FORK-FRIENDLY:** Make it easier for users to quickly coordinate on and execute a fork. This makes the payoff of capturing governance smaller.

The Uniswap case is particularly interesting: it's an intended behavior that the on-chain governance funds teams, which may develop future versions of the Uniswap protocol, but it's up to users to *opt in* to upgrading to those versions. This is a hybrid of on-chain and off-chain governance that leaves only a limited role for the on-chain side.

But limited governance is not an acceptable solution by itself; those areas where governance is needed the most (e.g., funds distribution for public goods) are themselves among the most vulnerable to attack. Public-goods funding is so vulnerable to attack because there is a very direct way for an attacker to profit from bad decisions: they can try to push through a bad decision that sends funds to themselves. Hence, we also need techniques to improve governance itself . . .

SOLUTION 2: NON-COIN-DRIVEN GOVERNANCE

A second approach is to use forms of governance that are not coin-voting-driven. But if coins do not determine what weight an account has in governance, what does? There are two natural alternatives:

1. **PROOF OF PERSONHOOD:** Systems that verify that accounts correspond to unique individual humans, so that governance can assign one vote per human. See Proof of Humanity and BrightID for two attempts to implement this.

2. **PROOF OF PARTICIPATION:** Systems that attest to the fact that some account corresponds to a person that has participated in some event, passed some educational training, or performed some useful work in the ecosystem. See POAP for one attempt to implement this.

There are also hybrid possibilities: one example is quadratic voting, which makes the power of a single voter proportional to the square root of the economic resources that they commit to a decision. Preventing people from gaming the system by splitting their resources across many identities requires proof of personhood, and the still-existent financial component allows participants to credibly signal how strongly they care about an issue, as well as how strongly they care about the ecosystem. Gitcoin quadratic funding is a form of quadratic voting, and quadratic-voting DAOs are being built.

Proof of participation is less well-understood. The key challenge is that determining what counts as how much participation itself requires a quite robust governance structure. It's possible that the easiest solution involves bootstrapping the system with a hand-picked choice of ten to one hundred early contributors, and then decentralizing over time as the selected participants of round n determine participation criteria for round $n + 1$. The possibility of a fork helps provide a path to recovery from, and an incentive against, governance going off the rails.

Proof of personhood and proof of participation both require some form of anti-collusion to ensure that the non-money resource being used to measure voting power remains non-financial, and does not itself end up inside of smart contracts that sell the governance power to the highest bidder.

SOLUTION 3: SKIN IN THE GAME

The third approach is to break the tragedy of the commons, by changing the rules of the vote itself. **Coin voting fails because while voters are *collectively* accountable for their decisions (if everyone votes for a terrible decision, everyone's coins drop to zero), each voter is not *individually* accountable (if a terrible decision happens, those who supported it suffer no more** than those who opposed it). **Can we make a voting system that changes this dynamic, and makes voters individually, and not just collectively, responsible for their decisions?**

Fork-friendliness is arguably a skin-in-the-game strategy, if forks are done in the way that Hive forked from Steem. In the case that a ruinous governance decision succeeds and can no longer be opposed inside the protocol, users can take it upon themselves to make a fork. Furthermore, in that fork, the coins that voted for the bad decision can be destroyed.

This sounds harsh, and perhaps it even feels like a violation of an implicit norm that the "immutability of the ledger" should remain sacrosanct when forking a coin. But the idea seems much more reasonable when seen from a different perspective. We keep the idea of a strong firewall where individual coin balances are expected to be inviolate, *but only apply that protection to coins that do not participate in governance.* If you participate in governance, even indirectly by putting your coins into a wrapper mechanism, then you may be held liable for the costs of your actions.

This creates individual responsibility: if an attack happens, and your coins vote for the attack, then your coins are destroyed. If your coins do not vote for the attack, your coins are safe. The responsibility propagates upward: if you put your coins into a wrapper contract and the wrapper contract votes for an attack, the wrapper contract's balance is wiped and so you lose your coins. If an attacker borrows XYZ from a DeFi lending platform, when the platform forks, anyone who lent XYZ loses out (note that this makes lending the governance token in general very risky; this is an intended consequence).

SKIN IN THE GAME IN DAY-TO-DAY VOTING

But the above only works for guarding against decisions that are truly extreme. What about smaller-scale heists, which unfairly favor attackers manipulating the economics of the governance but not severely enough to be ruinous? And what about, in the absence of any attackers at all, simple laziness, and the fact that coin-voting governance has no selection pressure in favor of higher-quality

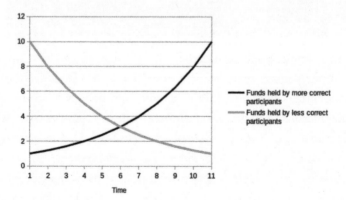

"Pure" futarchy has proven difficult to introduce, because in practice objective functions are very difficult to define (it's not just coin price that people want!), but various hybrid forms of futarchy may well work. Examples of hybrid futarchy include:

□ **VOTES AS BUY ORDERS:** Voting in favor of a proposal requires making an enforceable buy order to buy additional tokens at a price somewhat lower than the token's current price. This ensures that if a terrible decision succeeds, those who support it may be forced to buy their opponents out, but it also ensures that in more "normal" decisions coin holders have more slack to decide according to non-price criteria if they so wish.

□ **RETROACTIVE PUBLIC-GOODS FUNDING:** Public goods are funded by some voting mechanism *retroactively*, after they have already achieved a result. Users can buy *project tokens* to fund their project while signaling confidence in it; buyers of project tokens get a share of the reward if that project is deemed to have achieved a desired goal.

□ **ESCALATION GAMES:** Value-alignment on lower-level decisions is incentivized by the possibility to appeal to a higher-effort but higher-accuracy, higher-level process; voters whose votes agree with the ultimate decision are rewarded.

In the latter two cases, hybrid futarchy depends on some form of non-futarchy governance to measure against the objective function or serve as a dispute layer of last resort. However, this non-futarchy governance has several advantages that it does not if used directly: (i) it activates later, so it has access to more information, (ii) it is used less frequently, so it can expend less effort, and (iii) each use of it has greater consequences, so it's more acceptable to just rely on forking to align incentives for this final layer.

HYBRID SOLUTIONS

There are also solutions that combine elements of the above techniques. Some possible examples:

☐ **TIME DELAYS PLUS ELECTED-SPECIALIST GOVERNANCE:** This is one possible solution to the ancient conundrum of how to make a crypto-collateralized stablecoin whose locked funds can exceed the value of the profit-taking token without risking governance capture. The stablecoin uses a price oracle constructed from the median of values submitted by n (e.g., n = 13) elected providers. Coin voting chooses the providers, but it can only cycle out one provider each week. If users notice that coin voting is bringing in untrustworthy price providers, they have $n / 2$ weeks before the stablecoin breaks to switch to a different one.

☐ **FUTARCHY + ANTI-COLLUSION = REPUTATION:** Users vote with "reputation," a token that cannot be transferred. Users gain more reputation if their decisions lead to desired results, and lose reputation if their decisions lead to undesired results.

☐ **LOOSELY COUPLED (ADVISORY) COIN VOTES:** A coin vote does not directly implement a proposed change, instead it simply exists to make its outcome public, to build legitimacy for off-chain governance to implement that change. This can provide the benefits of coin votes, with fewer risks, as the legitimacy of a coin vote drops off automatically if evidence emerges that the coin vote was bribed or otherwise manipulated.

But these are all only a few possible examples. There is much more that can be done in researching and developing non-coin-driven governance algorithms. **The most important thing that can be done today is moving away from the idea that coin voting is the only legitimate form of governance decentralization.** Coin voting is attractive because it *feels* credibly neutral: anyone can go and get some units of the governance token on Uniswap. In practice, however, **coin voting may well only appear secure today precisely because of the imperfections in its neutrality** (namely, large portions of the supply staying in the hands of a tightly coordinated clique of insiders).

We should stay very wary of the idea that current forms of coin voting are "safe defaults." There is still much that remains to be seen about how they function under conditions of more economic stress and mature ecosystems and financial markets, and the time is now to start simultaneously experimenting with alternatives.

Special thanks to Karl Floersch, Dan Robinson, and Tina Zhen for feedback and review.

TRUST MODELS

vitalik.ca
August 20, 2021

One of the most valuable properties of many blockchain applications is *trustlessness*: the ability of the application to continue operating in an expected way without needing to rely on a specific actor to behave in a specific way even when their interests might change and push them to act in some different, unexpected way in the future. Blockchain applications are never *fully* trustless, but some applications are much closer to being trustless than others. If we want to make practical moves toward trust minimization, we want to have the ability to compare different degrees of trust.

First, my simple one-sentence definition of trust: **trust is the use of any assumptions about the behavior of other people**. If before the pandemic you would walk down the street without making sure to keep two meters' distance from strangers so that they could not suddenly take out a knife and stab you, that's a kind of trust: both trust that people are very rarely completely deranged, and trust that the people managing the legal system continue to provide strong incentives against that kind of behavior. When you run a piece of code written by someone else, you trust that they wrote the code honestly (whether due to their own sense of decency or due to an economic interest in maintaining their reputations), or at least that *there exist* enough people checking the code that a bug would be found. Not growing your own food is another kind of trust: trust that enough people will realize that it's in *their* interests to grow food so they can sell it to you. You can trust different sizes of groups of people, and there are different kinds of trust.

For the purposes of analyzing blockchain protocols, I tend to break down trust into four dimensions:

☐ How many people do you need to behave as you expect?

☐ Out of how many?

☐ What kinds of motivations are needed for those people to behave? Do they need to be altruistic, or just profit seeking? Do they need to be uncoordinated?

☐ How badly will the system fail if the assumptions are violated?

For now, let us focus on the first two. We can draw a graph:

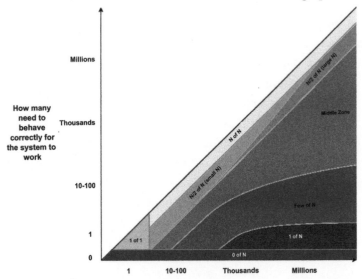

The darker the gray, the better. Let us explore the categories in more detail:

□ **1 OF 1:** There is exactly one actor, and the system works if (and only if) that one actor does what you expect them to. This is the traditional "centralized" model, and it is what we are trying to do better than.

□ **N OF N:** The "dystopian" world. You rely on a whole bunch of actors, *all* of whom need to act as expected for everything to work, with no backups if any of them fail.

□ **N/2 OF N:** This is how blockchains work—if the majority of the miners (or PoS validators) are honest. Notice that N/2 of N becomes significantly more valuable the larger the N gets; a blockchain with a few miners or validators dominating the network is much less interesting than a blockchain with its miners or validators widely distributed. That said, we want to improve on even this level of security, hence the concern around surviving 51% attacks.

□ **1 OF N:** There are many actors, and the system works as long as at least one of them does what you expect them to. Any system based on fraud proofs falls into this category, as do trusted setups though in that case the N is often smaller. Note that you do want the N to be as large as possible!

□ **FEW OF N:** There are many actors, and the system works as long as at least some small fixed number of them do what you expect them to do. Data availability checks fall into this category.

□ **0 OF N:** The system works as expected without any dependence whatsoever on external actors. Validating a block by checking it yourself falls into this category.

While all buckets other than "0 of N" can be considered "trust," they are very different from each other! Trusting that one particular person (or organization) will work as expected is very different from trusting that some *single person anywhere* will do what you expect them to. The "1 of N" model is arguably much closer to "0 of N" than it is to "N/2 of N" or "1 of 1." A "1 of N" model might perhaps feel like a "1 of 1" model because it feels like you're going through a single actor, but the reality of the two is very different: in a "1 of N" system, if the actor you're working with at the moment disappears or turns evil, you can just switch to another one, whereas in a "1 of 1" system you're screwed.

Particularly, note that even the correctness of the software you're running typically depends on a "few of N" trust model to ensure that if there are bugs in the code someone will catch them. With that fact in mind, trying really hard to go from "1 of N" to "0 of N" on some other aspect of an application is often like making a reinforced steel door for your house when the windows are open.

Another important question is: How does the system fail if your trust assumption is violated? In blockchains, the two most common types of failure are **liveness failure** and **safety failure**. A liveness failure is an event in which you are temporarily unable to do something you want to do (e.g., withdraw coins, get a transaction included in a block, read information from the blockchain). A safety failure is an event in which something actively happens that the system was meant to prevent (e.g., an invalid block gets included in a blockchain).

Here are a few examples of trust models of a few blockchain layer 2 protocols.* I use "**small N**" to refer to the set of partic-

---

* The models listed below are systems that rely on a "layer 1" blockchain like Ethereum or Bitcoin while providing it with greater capacity in some form or another.

ipants of the layer 2 system itself, and "**big N**" to refer to the participants of the blockchain; the assumption is always that the layer 2 protocol has a smaller community than the blockchain itself. I also limit my use of the word "liveness failure" to cases where coins are stuck for a significant amount of time; no longer being able to use the system but being able to near-instantly withdraw does not count as a liveness failure.

- □ **CHANNELS (INCLUDING STATE CHANNELS, LIGHTNING NETWORK):** "1 of 1" trust for liveness (your counterparty can temporarily freeze your funds, though the harms of this can be mitigated if you split coins between multiple counterparties); "N/2 of big-N" trust for safety (a blockchain 51% attack can steal your coins)

- □ **PLASMA (ASSUMING CENTRALIZED OPERATOR):** "1 of 1" trust for liveness (the operator can temporarily freeze your funds); "N/2 of big-N" trust for safety (blockchain 51% attack)

- □ **PLASMA (ASSUMING SEMI-DECENTRALIZED OPERATOR, E.G., DPOS):** "N/2 of small-N" trust for liveness; "N/2 of big-N" trust for safety

- □ **OPTIMISTIC ROLLUP:** "1 of 1" or "N/2 of small-N" trust for liveness (depends on operator type); "N/2 of big-N" trust for safety

- □ **ZK ROLLUP:** "1 of small-N" trust for liveness (if the operator fails to include your transaction, you can withdraw, and if the operator fails to include your withdrawal immediately, they cannot produce more batches and you can self-withdraw with the help of any full node of the rollup system); no safety-failure risks

- ☐ **ZK ROLLUP (WITH LIGHT-WITHDRAWAL ENHANCEMENT):** no liveness-failure risks; no safety-failure risks

Finally, there is the question of incentives: Does the actor you're trusting need to be very altruistic to act as expected or only slightly altruistic, or is being rational enough? Searching for fraud proofs is "by default" slightly altruistic, though just how altruistic it is depends on the complexity of the computation, and there are ways to modify the game to make it rational.

Assisting others with withdrawing from a ZK rollup is rational if we add a way to micro-pay for the service, so there is *really* little cause for concern that you won't be able to exit from a rollup with any significant use. Meanwhile, the greater risks of the other systems can be alleviated if we agree as a community to not accept 51% attack chains that revert too far in history or censor blocks for too long.

Conclusion: when someone says that a system "depends on trust," ask them in more detail what they mean! Do they mean "1 of 1," or "1 of N," or "N/2 of N"? Are they demanding these participants be altruistic or just rational? If altruistic, is it a tiny expense or a huge expense? And what if the assumption is violated—do you just need to wait a few hours or days, or do you have assets that are stuck forever? Depending on the answers, your own answer to whether or not you want to use that system might be very different.

CRYPTO CITIES

vitalik.ca
October 31, 2021

One interesting trend of the last year has been the growth of interest in local government, and in the idea of local governments that have wider variance and do more experimentation. Over the past year, Miami mayor Francis Suarez has pursued a tech-startup-like strategy of attracting interest in the city, frequently engaging with the mainstream tech industry and crypto community on Twitter. Wyoming now has a DAO-friendly legal structure, Colorado is experimenting with quadratic voting, and we're seeing more and more experiments making pedestrian-friendly street environments for the offline world. We're even seeing projects with varying degrees of radicalness—Culdesac, Telosa, CityDAO, Nkwashi, Prospera, and many more—trying to create entire neighborhoods and cities from scratch.

Another interesting trend of the last year has been the rapid mainstreaming of crypto ideas such as coins, non-fungible tokens, and decentralized autonomous organizations (DAOs). So what would happen if we combine the two trends together? Does it make sense to have a city with a coin, an NFT, a DAO, some record-keeping on-chain for anti-corruption, or even all four? As it turns out, there are already people trying to do just that:

☐ **CityCoins.co**, a project that sets up coins intended to become local media of exchange, where a portion of the issuance of the coin goes to the city government. Miami-Coin already exists, and San Francisco Coin appears to be coming soon.

☐ **Experiments with NFTs**, often as a way of funding local artists. Busan is hosting a government-backed conference exploring what they could do with NFTs.

☐ **Reno mayor Hillary Schieve's expansive vision for blockchain-ifying the city**, including NFT sales to support local art, a RenoDAO with RenoCoins issued to local residents that could get revenue from the government renting out properties, blockchain-secured lotteries, blockchain voting, and more.

☐ Much more ambitious projects **creating crypto-oriented cities from scratch**: see CityDAO, which describes itself as, well, "building a city on the Ethereum blockchain"—DAOified governance and all.

But are these projects, in their current form, good ideas? Are there any changes that could make them into *better* ideas? Let us find out . . .

WHY SHOULD WE CARE ABOUT CITIES?

Many national governments around the world are showing themselves to be inefficient and slow-moving in response to long-running problems and rapid changes in people's underlying needs. In short, many national governments are missing live players. Even worse, many of the outside-the-box political ideas that *are* being considered or implemented for national governance today are honestly quite terrifying. Do *you* want the USA to be taken over by a clone of the World War II-era Portuguese dictator António Salazar, or perhaps an "American Caesar," to beat down the evil scourge of American leftism? For every idea that can be reasonably described as freedom-expanding or democratic, there are ten that are just different forms of centralized control and walls and universal surveillance.

Now consider local governments. **Cities and states, as we've seen from the examples at the start of this post, are, at least in theory, capable of genuine dynamism.** There are large and very real differences of culture between cities, so it's easier to find a single city where there is public interest in adopting any particular radical idea than it is to convince an entire country to accept it. There are very real challenges and opportunities in local public goods, urban planning, transportation, and many other sectors in the governance of cities that could be addressed. Cities have tightly cohesive internal economies where things like widespread cryptocurrency adoption could realistically independently happen. Furthermore, it's less likely that experiments within cities will lead to terrible outcomes both because cities are regulated by higher-level governments and because cities have an escape valve: people who are unhappy with what's going on can more easily exit.

So all in all, it seems like the local level of government is a very undervalued one. And given that criticism of existing smart-city initiatives often heavily focuses on concerns around centralized governance, lack of transparency, and data privacy, blockchain and cryptographic technologies seem like a promising key ingredient for a more open and participatory way forward.

WHAT ARE CITY PROJECTS UP TO TODAY?

Quite a lot actually! Each of these experiments is still small scale and largely still trying to find its way around, but they are all at least seeds that could turn into interesting things. Many of the most advanced projects are in the United States, but there is interest across the world; over in Korea the government of Busan is running an NFT conference. Here are a few examples of what is being done today.

BLOCKCHAIN EXPERIMENTS IN RENO

Reno, Nevada, mayor Hillary Schieve is a blockchain fan, focusing primarily on the Tezos ecosystem, and she has recently been exploring blockchain-related ideas in the governance of her city:

☐ **Selling NFTs to fund local art**, starting with an NFT of the "Space Whale" sculpture in the middle of the city.

□ **Creating a RenoDAO**, governed by Reno coins that residents would be eligible to receive via an airdrop. The RenoDAO could start to get sources of revenue; one proposed idea was for the city to rent out properties that it owns and use the revenue to fund a DAO.

□ **Using blockchains to secure all kinds of processes**; for example, blockchain-secured random number generators for casinos, blockchain-secured voting, etc.

CITYCOINS.CO

CityCoins.co is a project built on Stacks, a blockchain run by an unusual "proof of transfer" (for some reason abbreviated PoX, not PoT) block-production algorithm that is built around the Bitcoin blockchain and ecosystem. Seventy percent of the coin's supply is generated by an ongoing sale mechanism: anyone with STX (the Stacks native token) can send their STX to the city-coin contract to generate city coins; the STX revenues are distributed to existing city-coin holders who stake their coins. The remaining 30% is made available to the city government.

CityCoins has made the interesting decision of trying to make an economic model that does not depend on any government support. The local government does not need to be involved in creating a CityCoins.co coin; a community group can launch a coin by themselves. An FAQ-provided answer to "What can I do with CityCoins?" includes examples like "CityCoins communities will create apps that use tokens for rewards" and "local businesses can provide discounts or benefits to people who . . . stack their CityCoins." In practice, however, the MiamiCoin community is not going at it alone; the Miami government has already de facto publicly endorsed it.

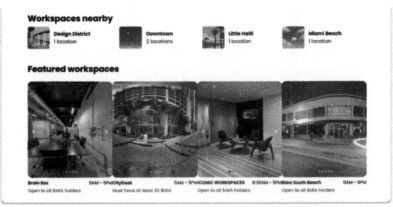

MiamiCoin hackathon winner: a site that allows coworking spaces to give preferential offers to MiamiCoin holders.

CITYDAO

CityDAO is the most radical of the experiments: Unlike Miami and Reno, which are existing cities with existing infrastructure to be upgraded and people to be convinced, CityDAO is a DAO with legal status under the Wyoming DAO law trying to create entirely new cities from scratch.

So far, the project is still in its early stages. The team is currently finalizing a purchase of their first plot of land in a far-off corner of Wyoming. The plan is to start with this plot of land, and then add other plots of land in the future, to build cities that are governed by a DAO and make heavy use of radical economic ideas like Harberger taxes to allocate the land, make collective decisions, and manage resources. Their DAO is one of the progressive few that is avoiding coin-voting governance; instead, the governance is a voting scheme based on "citizen" NFTs, and ideas have been floated to further limit votes to one per person by using Proof of Humanity verification. The NFTs are currently being sold to crowdfund the project; you can buy them on OpenSea.

WHAT DO I THINK CITIES COULD BE UP TO?

Obviously there are a lot of things that cities could do in principle. They could add more bike lanes, they could use CO_2 meters and far-UVC light to more effectively reduce COVID spread without inconveniencing people, and they could even fund life-extension research. But my primary specialty is blockchains and this post is about blockchains, so . . . let's focus on blockchains.

I would argue that there are two distinct categories of blockchain ideas that make sense:

1. Using blockchains to create **more trusted, transparent, and verifiable versions of existing processes**.

2. Using blockchains to implement **new and experimental forms of ownership** for land and other scarce assets, as well as **new and experimental forms of democratic governance**.

There's a natural fit between blockchains and both of these

categories. Anything happening on a blockchain is very easy to publicly verify, with lots of ready-made, freely available tools to help people do that. Any application built on a blockchain can immediately plug in to and interface with other applications in the entire global blockchain ecosystem. Blockchain-based systems are efficient in a way that paper is not, and publicly verifiable in a way that centralized computing systems are not—a necessary combination if you want to, say, make a new form of voting that allows citizens to give high-volume real-time feedback on hundreds or thousands of different issues.

So let's get into the specifics.

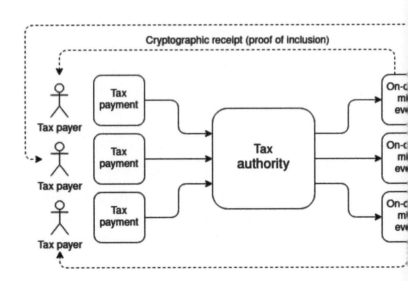

WHAT ARE SOME EXISTING PROCESSES THAT BLOCKCHAINS COULD MAKE MORE TRUSTED AND TRANSPARENT?

One simple idea that plenty of people, including government officials around the world, have brought up to me on many occasions is the idea of governments creating a white-listed internal-use-only stablecoin for tracking internal government payments. Every tax payment from an individual or organization could be tied to a publicly visible on-chain record minting that number of coins (if we want individual tax payment quantities to be private, there are zero-knowledge ways to make only the total public but still convince everyone that it was computed correctly). Transfers between departments could be done "in the clear," and the coins would be redeemed only by individual contractors or employees claiming their payments and salaries.

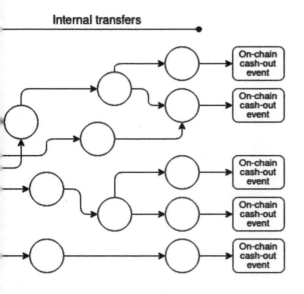

Internal transfers

This system could easily be extended. For example, procurement processes for choosing which bidder wins a government contract could largely be done on-chain.

Many more processes could be made more trustworthy with blockchains:

☐ **FAIR RANDOM NUMBER GENERATORS (E.G., FOR LOTTERIES)—** VDFs, such as the one Ethereum is expected to include, could serve as a fair random number generator that could be used to make government-run lotteries more trustworthy. Fair randomness could also be used for many other use cases, such as sortition as a form of government.

☐ **CERTIFICATES**—for example, cryptographic proofs that some particular individual is a resident of the city—could be done on-chain for added verifiability and security (e.g., if such certificates are issued on-chain, it would become obvious if a large number of false certificates are issued). This can be used by all kinds of local-government-issued certificates.

☐ **ASSET REGISTRIES**, for land and other assets, as well as more complicated forms of property ownership such as development rights. Due to the need for courts to be able to make assignments in exceptional situations, these registries will likely never be fully decentralized bearer instruments in the same way that cryptocurrencies are, but putting records on-chain can still make it easier to see what happened in what order in a dispute.

Eventually, even **voting** could be done on-chain. Here, many complexities and dragons loom, and it's really important to be careful; a sophisticated solution combining blockchains, zero-knowledge proofs, and other cryptography is needed to achieve all the desired privacy and security properties. However, if humanity is ever going to move to electronic voting at all, local government seems like the perfect place to start.

WHAT ARE SOME RADICAL ECONOMIC AND GOVERNANCE EXPERIMENTS THAT COULD BE INTERESTING?

But in addition to these kinds of blockchain overlays onto things that governments *already* do, we can also look at blockchains as an opportunity for governments to make completely *new* and radical experiments in economics and governance. These are not necessarily final ideas on what I think should be done; they are initial explorations and suggestions for possible directions. Once an experiment starts, real-world feedback is often by far the most useful variable to determine how the experiment should be adjusted in the future.

EXPERIMENT #1: A MORE COMPREHENSIVE VISION OF CITY TOKENS

CityCoins.co is one vision for how city tokens could work. But it is far from the only vision. Indeed, the CityCoins.co approach has significant risks, particularly in how the economic model is heavily tilted toward early adopters. Seventy percent of the STX revenue from minting new coins is given to *existing stakers of the city coin*. More coins will be issued in the next five years than in the fifty years that follow. It's a good deal for the government in 2021, but what about 2051? Once a government endorses a particular city coin, it becomes difficult for it to change directions in the future. Hence, it's important for city governments to think carefully about these issues, and choose a path that makes sense for the long term.

Here is a different possible sketch of a narrative of how city tokens might work. It's far from the *only* possible alternative to the CityCoins.co vision. In any case, city tokens are a wide design space, and there are many different options worth considering. Anyway, here goes . . .

The concept of home ownership in its current form is a notable double-edged sword, and the specific ways in which it's actively encouraged and legally structured is considered by many to be one of the biggest economic policy mistakes that we are making today. **There is an inevitable political tension between a home as a place to live and a home as an investment asset**, and the pressure to satisfy communities who care about the latter often ends up severely harming the affordability of the former. Residents in a city either own a home, making them massively over-exposed to land prices and introducing perverse incentives to fight against construction of new homes, or rent a home, making them *negatively* exposed to the real estate market and thus putting them economically at odds with the goal of making a city a nice place to live.

But even despite all of these problems, many still find home ownership to be not just a good personal choice, but something worthy of actively subsidizing or socially encouraging. One big reason is that it nudges people to save money and build up their net worth. Another big reason is that despite its flaws, it creates economic alignment between residents and the communities they live in. **But what if we could give people a way to save and create that economic alignment without the flaws?** What if we could create a divisible and fungible city token, that residents could hold as many units of as they can afford or feel comfortable with, and whose value goes up as the city prospers?

First, let's start with some possible objectives. Not all are necessary; a token that accomplishes only three of the five is already a big step forward. But we'll try to hit as many of them as possible:

□ **GET SUSTAINABLE SOURCES OF REVENUE FOR THE GOVERNMENT:** The city token economic model should avoid redirecting *existing* tax revenue; instead, it should find *new* sources of revenue.

□ **CREATE ECONOMIC ALIGNMENT BETWEEN RESIDENTS AND THE CITY:** This means first of all that the coin itself should clearly become more valuable as the city becomes more attractive. But it also means that the economics should actively encourage *residents* to hold the coin more than faraway hedge funds.

□ **PROMOTE SAVING AND WEALTH-BUILDING:** Home ownership does this—as home owners make mortgage payments, they build up their net worth by default. City tokens could do this too, making it attractive to accumulate coins over time, and even gamifying the experience.

□ **ENCOURAGE MORE PRO-SOCIAL ACTIVITY:** Such as positive actions that help the city and more sustainable use of resources.

□ **BE EGALITARIAN:** Don't unduly favor wealthy people over poor people (as badly designed economic mechanisms often do accidentally). A token's divisibility, avoiding a sharp binary divide between haves and have-nots, does a lot already, but we can go further—for example, by allocating a large portion of new issuance to residents as a UBI.*

One pattern that seems to easily meet the first three objectives is providing benefits to holders: If you hold at least x coins (where x can go up over time), you get some set of services for free. MiamiCoin is trying to encourage businesses to do this,

---

* Universal basic income, in which all residents would receive an equal, unconditional income at regular intervals.

but we could go further and make *government* services work this way too. One simple example would be making existing public parking spaces only available for free to those who hold at least some number of coins in a locked-up form. This would serve a few goals at the same time:

☐ Create an **incentive to hold the coin**, sustaining its value.

☐ Create an **incentive specifically for *residents* to hold the coin**, as opposed to otherwise-unaligned faraway investors. Furthermore, the incentive's usefulness is capped per person, so it encourages widely distributed holdings.

☐ Creates **economic alignment** (city becomes more attractive → more people want to park → coins have more value). **Unlike home ownership, this creates alignment with an *entire town*,** and not merely a very specific location in a town.

☐ **Encourage sustainable use of resources** by reducing usage of parking spots (though people without coins who really need them could still pay), supporting many local governments' desires to open up more pedestrian-friendly space on the roads. Alternatively, restaurants could also be allowed to lock up coins through the same mechanism and claim parking spaces to use for outdoor seating.

But to avoid perverse incentives, it's extremely important to avoid overly depending on one specific idea and instead to have a diverse array of possible revenue sources. **One excellent gold mine of places to give city tokens value, and at the same time experiment with novel governance ideas, is zoning.** If you hold at least y coins, then you can quadratically vote on the fee that nearby landowners have to pay to bypass zoning restrictions. This hybrid market- plus direct-democracy-based approach would be much more efficient than current overly cumbersome permitting processes, and the fee itself would be another source of government revenue. More generally, any of the ideas in the next section could be combined with city tokens to give city-token holders more places to use them.

EXPERIMENT #2: MORE RADICAL AND PARTICIPATORY FORMS OF GOVERNANCE

This is where *Radical Markets** ideas such as Harberger taxes, quadratic voting, and quadratic funding come in. I already brought up some of these ideas in the section above, but you don't have to have a dedicated city token to do them. Some limited government use of quadratic voting and funding has already happened: see the Colorado Democratic Party and the Taiwanese presidential hackathon, as well as not-yet-government-backed experiments like Gitcoin's Boulder Downtown Stimulus. But we could do more!

One obvious place where these ideas can have long-term value is giving developers incentives to improve the **aesthetics of buildings**. Harberger taxes and other mechanisms could be used to radically reform **zoning** rules, and blockchains could be used to administer such mechanisms in a more trustworthy and efficient way. Another idea that is more viable in the short term is **subsidizing local businesses**, similar to the Downtown Stimulus but on a larger and more permanent scale. Businesses produce various kinds of positive externalities in their local communities all the time, and those externalities could be more effectively

---

* Again, referencing the book of this name (and its family of concepts) by Eric Posner and E. Glen Weyl.

rewarded. **Local news** could be quadratically funded, revitalizing a long-struggling industry. Pricing for **advertisements** could be set based on real-time votes of how much people enjoy looking at each particular ad, encouraging more originality and creativity.

More democratic feedback (and possibly even retroactive democratic feedback!) could plausibly create better incentives in all of these areas. And **twenty-first-century digital democracy through real-time online quadratic voting and funding could plausibly do a much better job than twentieth-century democracy, which seems in practice to have been largely characterized by rigid building codes and obstruction at planning and permitting hearings**. And of course, if you're going to use blockchains to secure voting, starting off by doing it with fancy new kinds of votes seems far more safe and politically feasible than re-fitting existing voting systems.

Mandatory solarpunk picture intended to evoke a positive image of what might happen to our cities if real-time quadratic votes could set subsidies and prices for everything.

CONCLUSIONS

There are a lot of worthwhile ideas for cities to experiment with that could be attempted by existing cities or by new cities. New cities of course have the advantage of not having existing residents with existing expectations of how things should be done; but the concept of creating a new city itself is, in modern times, relatively untested. Perhaps the multibillion-dollar capital pools in the hands of people and projects enthusiastic to try new things could get us over the hump. But even then, existing cities will likely continue to be the place where most people live for the foreseeable future, and existing cities can use these ideas too.

Blockchains can be very useful in both the more incremental and more radical ideas that were proposed here, even despite the inherently "trusted" nature of a city government. Running any new or existing mechanism on-chain gives the public an easy ability to verify that everything is following the rules. Public chains are better: the benefits from existing infrastructure for users to independently verify what is going on far outweigh the losses from transaction fees, which are expected to quickly decrease very soon from rollups and sharding. If strong privacy is required, blockchains can be combined with zero-knowledge cryptography to give privacy and security at the same time.

The main trap that governments should avoid is too quickly sacrificing optionality. An *existing* city could fall into this trap by launching a bad city token instead of taking things more slowly and launching a good one. A new city could fall into this trap by selling off too much land, sacrificing the entire upside to a small group of early adopters. Starting with self-contained experiments, and taking things slowly on moves that are truly irreversible, is ideal. But at the same time, it's also important to seize the opportunity in the first place. There's a lot that can and should be improved with cities, and a lot of opportunities; despite the challenges, crypto cities broadly are an idea whose time has come.

Special thanks to Mr. Silly and Tina Zhen for early feedback on the post, and to a long list of people for discussion of the ideas.

SOULBOUND

vitalik.ca
January 26, 2022

One feature of *World of Warcraft* that is second nature to its players, but goes mostly undiscussed outside of gaming circles, is the concept of *soulbound* items. A soulbound item, once picked up, cannot be transferred or sold to another player.

Most very powerful items in the game are soulbound, and typically require completing a complicated quest or killing a very powerful monster, usually with the help of anywhere from four to thirty-nine other players. Hence, in order to get your character anywhere close to having the best weapons and armor, you have no choice but to participate in killing some of these extremely difficult monsters yourself.

The purpose of the mechanism is fairly clear: it keeps the game challenging and interesting, by making sure that to get the best items you have to actually go and do the hard thing and figure out how to kill the dragon. You can't just go kill boars ten hours a day for a year, get thousands of gold, and buy the epic magic armor from other players who killed the dragon for you.

Of course, the system is very imperfect: you could just pay a team of professionals to kill the dragon with you and let you collect the loot, or even outright buy a character on a secondary market, and do this all with out-of-game US dollars so you don't even have to kill boars. But even still, it makes for a much better game than one in which every item always has a price.

WHAT IF NFTS COULD BE SOULBOUND?

NFTs in their current form have many of the same properties as rare and epic items in a massively multiplayer online game. They have social signaling value: people who have them can show them off, and there are more and more tools precisely to help users do that. Very recently, Twitter started rolling out an integration that allows users to show off their NFTs on their picture profile.

But what exactly are these NFTs signaling? Certainly, one part of the answer is some kind of skill in acquiring NFTs and knowing which NFTs to acquire. But because NFTs are tradable items, another big part of the answer inevitably becomes that NFTs are about signaling wealth.

If someone shows you that they have an NFT that is obtainable by doing X, you can't tell whether they did X themselves or whether they just paid someone else to do X. Some of the time this is not a problem: for an NFT supporting a charity, someone buying it off the secondary market is sacrificing their own funds for the cause and they are helping the charity by contributing to

CryptoPunks are now regularly being sold for many millions of dollars, and they are not even the most expensive NFTs out there.

others' incentive to buy the NFT, and so there is no reason to discriminate against them. And indeed, a lot of good can come from charity NFTs alone. But what if we want to create NFTs that are not just about who has the most money, and that actually try to signal something else?

Perhaps the best example of a project trying to do this is POAP, the "proof of attendance protocol." POAP is a standard by which projects can send NFTs that represent the idea that the recipient personally participated in some event.

POAP is an excellent example of an NFT that works better if it could be soulbound. If someone is looking at your POAP, they are not interested in whether or not you paid someone who attended some event. They are interested in whether or not *you personally* attended that event. Proposals to put certificates (e.g., driver's licenses, university degrees, proof of age) on-chain face a similar problem: they would be much less valuable if someone who doesn't meet the condition themselves could just go buy one from someone who does.

While transferable NFTs have their place and can be really

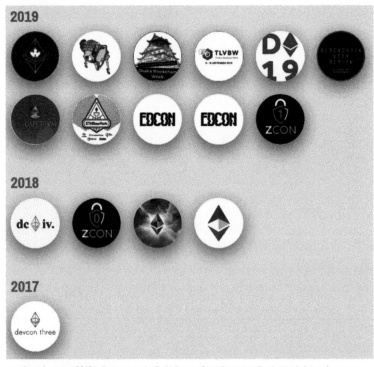

Part of my own POAP collection, much of which came from the events that I attended over the years.

valuable on their own for supporting artists and charities, there is also a large and under-explored design space of what *non-transferable* NFTs could become.

WHAT IF GOVERNANCE RIGHTS COULD BE SOULBOUND?

This is a topic I have written about ad nauseam, but it continues to be worth repeating: **there are very bad things that can easily happen to governance mechanisms if governance power is easily transferable**. This is true for two primary types of reasons:

□ If the goal is for governance power *to be widely distributed*, then transferability is counterproductive as concentrated interests are more likely to buy the governance rights up from everyone else.

□ If the goal is for governance power *to go to the competent*, then transferability is counterproductive because nothing stops the governance rights from being bought up by the determined but incompetent.

If you take the proverb that "those who most want to rule people are those least suited to do it" seriously, then you should be suspicious of transferability, precisely because transferability makes governance power flow away from the meek who are most likely to provide valuable input to governance and toward the power-hungry who are most likely to cause problems.

So what if we try to make governance rights non-transferable? What if we try to make a CityDAO where more voting power goes to the people who actually live in the city, or at least is reliably democratic and avoids undue influence by whales hoarding a large number of citizen NFTs? What if DAO governance of blockchain protocols could somehow make governance power conditional on participation? Once again, a large and fruitful design space opens up that today is difficult to access.

IMPLEMENTING NON-TRANSFERABILITY IN PRACTICE

POAP has made the technical decision to not block transfer-ability of the POAPs themselves. There are good reasons for this: users might have a good reason to want to migrate all their assets from one wallet to another (e.g., for security), and the security of non-transferability implemented "naïvely" is not very strong anyway because users could just create a wrapper account that holds the NFT and then sell the ownership of that.

And indeed, there have been quite a few cases where POAPs have frequently been bought and sold when an economic rationale was there to do so. Adidas recently released a POAP for free to their fans that could give users priority access at a merchandise sale. What happened? Well, of course, many of the POAPs were quickly transferred to the highest bidder.

More transfers than items. And not the only time.

To solve this problem, the POAP team is suggesting that developers who care about non-transferability implement checks on their own: they could check on-chain if the current owner is the same address as the original owner, and they could add more sophisticated checks over time if deemed necessary. This is, for now, a more future-proof approach.

Perhaps the one NFT that is the most robustly non-transferable today is the Proof of Humanity attestation.* Theoretically, anyone can

---

* Proof of Humanity is a project designed to establish unique human identities on a blockchain without relying on central authorities such as governments or corporations. It is used by other crypto projects that need to confirm the personhood of participants.

create a Proof of Humanity profile with a smart-contract account that has transferable ownership, and then sell that account. But the Proof of Humanity protocol has a revocation feature that allows the original owner to make a video asking for a profile to be removed, and a Kleros court decides whether or not the video was from the same person as the original creator. Once the profile is successfully removed, they can reapply to make a new profile. Hence, if you buy someone else's Proof of Humanity profile, your possession can be very quickly taken away from you, making transfers of ownership nonviable. Proof of Humanity profiles are de facto soulbound, and infrastructure built on top of them could allow for on-chain items in general to be soulbound to particular humans.

Can we limit transferability without going all the way and basing everything on Proof of Humanity? It becomes harder, but there are medium-strength approaches that are probably good enough for some use cases. Making an NFT bound to an ENS name is one simple option, if we assume that users care enough about their ENS names that they are not willing to transfer them. For now, what we're likely to see is a spectrum of approaches to limit transferability, with different projects choosing different tradeoffs between security and convenience.

NON-TRANSFERABILITY AND PRIVACY

Cryptographically strong privacy for transferable assets is fairly easy to understand: you take your coins, put them into tornado.cash* or a similar platform, and withdraw them into a fresh account. But how can we add privacy for soulbound items where you cannot

---

* Whereas normally a blockchain like Ethereum publishes the senders and recipients of all transactions, Tornado Cash is a protocol that enables private transactions by masking the link between sender and receiver.

just move them into a fresh account or even a smart contract? If Proof of Humanity starts getting more adoption, privacy becomes even more important, as the alternative is all of our activity being mapped on-chain directly to a human face.

Fortunately, a few fairly simple technical options are possible:

□ Store the item at an address which is the hash of (i) an index, (ii) the recipient address, and (iii) a secret belonging to the recipient. You could reveal your secret to an interface that would then scan for all possible items that belong to your, but no one without your secret could see which items are yours.

□ Publish a hash of a bunch of items, and give each recipient their Merkle branch.*

□ If a *smart contract* needs to check whether you have an item of some type, you can provide a ZK-SNARK.**

Transfers could be done on-chain; the simplest technique may just be a transaction that calls a factory contract to make the old item invalid and the new item valid, using a ZK-SNARK to prove that the operation is valid.

Privacy is an important part of making this kind of ecosystem work well. In some cases, the underlying thing that the item is representing is already public, and so there is no point in trying to add privacy. But in many other cases, users would not want to reveal everything that they have. If, one day in the future, being

---

* Merkle trees are a cryptographic technique, central to the design of Ethereum, used to verify that a set of data has not been tampered with. A Merkle branch is part of such a tree.
** ZK-SNARK stands for "Zero-Knowledge Succinct Non-Interactive Argument of Knowledge." It is a technique for providing cryptographic evidence that a party holds certain information without revealing what that information is.

vaccinated becomes a POAP, one of the worst things we could do would be to create a system where the POAP is automatically advertised for everyone to see and everyone has no choice but to let their medical decision be influenced by what would look cool in their particular social circle. Privacy being a core part of the design can avoid these bad outcomes and increase the chance that we create something great.

FROM HERE TO THERE

A common criticism of the "web3" space as it exists today is how money-oriented everything is. People celebrate the ownership, and outright waste, of large amounts of wealth, and this limits the appeal and the long-term sustainability of the culture that emerges around these digital collectibles. There are of course important benefits that even financialized NFTs can provide, such as funding artists and charities that would otherwise go unrecognized. However, there are limits to that approach, and a lot of under-explored opportunity in trying to go beyond financialization. Making more items in the crypto space "soulbound" can be one path toward an alternative, where NFTs can represent much more of who you are and not just what you can afford.

However, there are technical challenges to doing this, and an uneasy "interface" between the desire to limit or prevent transfers and a blockchain ecosystem where so far all of the standards are designed around maximum transferability. Attaching items to "identity objects" that users are either unable (as with Proof of Humanity profiles) or unwilling (as with ENS names) to trade away seems like the most promising path, but challenges remain in making this easy to use, private, and secure. We need more effort on thinking through and solving these challenges. If we can, this opens a much wider door to blockchains being at the center of ecosystems that are collaborative and fun, and not just about money.

APPENDIX

ETHEREUM WHITEPAPER: A NEXT-GENERATION SMART CONTRACT AND DECENTRALIZED APPLICATION PLATFORM

Satoshi Nakamoto's development of Bitcoin in 2009 has often been hailed as a radical development in money and currency, being the first example of a digital asset which simultaneously has no backing or "intrinsic value" and no centralized issuer or controller. However, another, arguably more important, part of the Bitcoin experiment is the underlying blockchain technology as a tool of distributed consensus, and attention is rapidly starting to shift to this other aspect of Bitcoin. Commonly cited alternative applications of blockchain technology include using on-blockchain digital assets to represent custom currencies and financial instruments ("colored coins"), the ownership of an underlying physical device ("smart property"), non-fungible assets such as domain names ("Namecoin"), as well as more complex applications involving having digital assets being directly controlled by a piece of code implementing arbitrary rules ("smart contracts") or even blockchain-based "decentralized autonomous organizations" (DAOs). What Ethereum intends to provide is a blockchain with a built-in fully fledged Turing-complete programming language that can be used to create "contracts" that can be used to encode arbitrary state transition functions, allowing users to create any of the systems described above, as well as many others that we have not yet imagined, simply by writing up the logic in a few lines of code.

INTRODUCTION TO BITCOIN AND EXISTING CONCEPTS

HISTORY

The concept of decentralized digital currency, as well as alternative applications like property registries, has been around for decades. The anonymous e-cash protocols of the 1980s and the 1990s, mostly reliant on a cryptographic primitive known as Chaumian blinding, provided a currency with a high degree of privacy, but the protocols largely failed to gain traction because of their reliance on a centralized intermediary. In 1998, Wei Dai's b-money became the first proposal to introduce the idea of creating money through solving computational puzzles as well as decentralized consensus, but the proposal was scant on details as to how decentralized consensus could actually be implemented. In 2005, Hal Finney introduced a concept of "reusable proofs of work," a system which uses ideas from b-money together with Adam Back's computationally difficult Hashcash puzzles to create a concept for a cryptocurrency, but once again fell short of the ideal by relying on trusted computing as a backend. In 2009, a decentralized currency was for the first time implemented in practice by Satoshi Nakamoto, combining established primitives for managing ownership through public key cryptography with a consensus algorithm for keeping track of who owns coins, known as "proof of work."

The mechanism behind proof of work was a breakthrough in the space because it simultaneously solved two problems. First, it provided a simple and moderately effective consensus algorithm, allowing nodes in the network to collectively agree on a set of canonical updates to the state of the Bitcoin ledger. Second, it provided a mechanism for allowing free entry into the consensus process, solving the political problem of deciding who gets to influence the consensus, while simultaneously preventing Sybil attacks. It does this by substituting a formal barrier to participation, such as the requirement to be registered as a unique entity on a particular list, with an economic barrier—the weight of a single node in the consensus voting process is directly proportional to the computing power that the node brings. Since then, an alternative approach has been proposed called *proof of stake*, calculating the weight of a node as being proportional to its currency holdings and not computational resources; the discussion of the relative merits of the two approaches is beyond the scope of this paper but it should be noted that both approaches can be used to serve as the backbone of a cryptocurrency.

BITCOIN AS A STATE TRANSITION SYSTEM

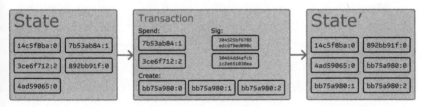

From a technical standpoint, the ledger of a cryptocurrency such as Bitcoin can be thought of as a state transition system, where there is a "state" consisting of the ownership status of all existing bitcoins and a "state transition function" that takes a state and a transaction and outputs a new state which is the result. In a standard banking system, for example, the state is a balance sheet, a transaction is a request to move $x from A to B, and the state transition function reduces the value in A's account by $x and increases the value in B's account by $x. If A's account has less than $x in the first place, the state transition function returns an error. Hence, one can formally define:

```
APPLY(S,TX) -> S' or ERROR
```

In the banking system defined above:

```
APPLY({ Alice: $50, Bob: $50 },"send $20 from Alice
to Bob") = { Alice: $30, Bob: $70 }
```

But:

```
APPLY({ Alice: $50, Bob: $50 },"send $70 from Alice
to Bob") = ERROR
```

The "state" in Bitcoin is the collection of all coins (technically, "unspent transaction outputs" or UTXO) that have been minted and not yet spent, with each UTXO having a denomination and an owner (defined by a twenty-byte address which is essentially a cryptographic public key*). A transaction contains one or more inputs, with each input containing a reference to an existing

---

* *In original:* A sophisticated reader may notice that, in fact, a Bitcoin address is the hash of the elliptic curve public key, and not the public key itself. However, it is, in fact, perfectly legitimate cryptographic terminology to refer to the pubkey hash as a public key itself. This is because Bitcoin's cryptography can be considered to be a custom digital signature algorithm, where the public key consists of the hash of the ECC pubkey, the signature consists of the ECC pubkey concatenated with the ECC signature, and the verification algorithm involves checking the ECC pubkey in the signature against the ECC pubkey hash provided as a public key and then verifying the ECC signature against the ECC pubkey.

UTXO and a cryptographic signature produced by the private key associated with the owner's address, and one or more outputs, with each output containing a new UTXO to be added to the state.

The state transition function `APPLY(S,TX) -> S'` can be defined roughly as follows:

1. For each input in `TX`:
    □   If the referenced UTXO is not in `S`, return an error.
    □   If the provided signature does not match the owner of the UTXO, return an error.

2. If the sum of the denominations of all input UTXO is less than the sum of the denominations of all output UTXO, return an error.

3. Return `S` with all input UTXO removed and all output UTXO added.

The first half of the first step prevents transaction senders from spending coins that do not exist, the second half of the first step prevents transaction senders from spending other people's coins, and the second step enforces conservation of value. In order to use this for payment, the protocol is as follows. Suppose Alice wants to send 11.7 BTC to Bob. First, Alice will look for a set of available UTXO that she owns that totals up to at least 11.7 BTC. Realistically, Alice will not be able to get exactly 11.7 BTC; say that the smallest she can get is $6 + 4 + 2 = 12$. She then creates a transaction with those three inputs and two outputs. The first output will be 11.7 BTC with Bob's address as its owner, and the second output will be the remaining 0.3 BTC "change," with the owner being Alice herself.

MINING

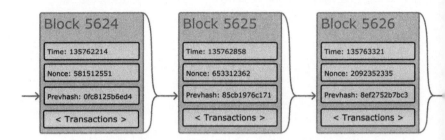

If we had access to a trustworthy centralized service, this system would be trivial to implement; it could simply be coded exactly as described, using a centralized server's hard drive to keep track of the state. However, with Bitcoin we are trying to build a decentralized currency system, so we will need to combine the state transaction system with a consensus system in order to ensure that everyone agrees on the order of transactions. Bitcoin's decentralized consensus process requires nodes in the network to continuously attempt to produce packages of transactions called "blocks." The network is intended to produce roughly one block every ten minutes, with each block containing a timestamp, a nonce, a reference to (i.e., hash of) the previous block and a list of all of the transactions that have taken place since the previous block. Over time, this creates a persistent, ever-growing "block-chain" that constantly updates to represent the latest state of the Bitcoin ledger.

The algorithm for checking if a block is valid, expressed in this paradigm, is as follows:

1. Check if the previous block referenced by the block exists and is valid.

2. Check that the timestamp of the block is greater than that of the previous block* and less than two hours into the future.

3. Check that the proof of work on the block is valid.

4. Let S[0] be the state at the end of the previous block.

5. Suppose TX is the block's transaction list with n transactions. For all i in 0...n-1, set S[i+1] = APPLY(S[i],TX-[i]). If any application returns an error, exit and return false.

6. Return true, and register S[n] as the state at the end of this block.

Essentially, each transaction in the block must provide a valid state transition from what was the canonical state before the transaction was executed to some new state. Note that the state is not encoded in the block in any way; it is purely an abstraction to be remembered by the validating node and can only be (securely) computed for any block by starting from the genesis state and sequentially applying every transaction in every block. Additionally, note that the order in which the miner includes transactions into the block matters; if there are two transactions A and B in a block such that B spends a UTXO created by A, then the block will be valid if A comes before B but not otherwise.

The one validity condition present in the above list that is not found in other systems is the requirement for "proof of work." The precise condition is that the double-SHA256 hash of every block, treated as a 256-bit number, must be less than a dynamically adjusted target, which as of the time of this writing is

---

* *In original:* Technically, the median of the eleven previous blocks.

approximately 2^{187}. The purpose of this is to make block creation computationally "hard," thereby preventing Sybil attackers from remaking the entire blockchain in their favor. Because SHA256 is designed to be a completely unpredictable pseudorandom function, the only way to create a valid block is simply trial and error, repeatedly incrementing the nonce and seeing if the new hash matches.

At the current target of $\sim2^{187}$, the network must make an average of $\sim2^{69}$ tries before a valid block is found; in general, the target is recalibrated by the network every 2,016 blocks so that on average a new block is produced by some node in the network every ten minutes. In order to compensate miners for this computational work, the miner of every block is entitled to include a transaction giving themselves 25 BTC out of nowhere. Additionally, if any transaction has a higher total denomination in its inputs than in its outputs, the difference also goes to the miner as a "transaction fee." Incidentally, this is also the only mechanism by which BTC are issued; the genesis state contained no coins at all.

In order to better understand the purpose of mining, let us examine what happens in the event of a malicious attacker. Since Bitcoin's underlying cryptography is known to be secure, the attacker will target the one part of the Bitcoin system that is not protected by cryptography directly: the order of transactions. The attacker's strategy is simple:

1. Send 100 BTC to a merchant in exchange for some product (preferably a rapid-delivery digital good).

2. Wait for the delivery of the product.

3. Produce another transaction sending the same 100 BTC to himself.

4. Try to convince the network that his transaction to himself was the one that came first.

Once step (1) has taken place, after a few minutes some miner will include the transaction in a block, say block number 270,000. After about one hour, five more blocks will have been added to the chain after that block, with each of those blocks indirectly pointing to the transaction and thus "confirming" it. At this point, the merchant will accept the payment as finalized and deliver the product; since we are assuming this is a digital good, delivery is instant. Now, the attacker creates another transaction sending the 100 BTC to himself. If the attacker simply releases it into the wild, the transaction will not be processed; miners will attempt to run APPLY(S,TX) and notice that TX consumes a UTXO which is no longer in the state. So instead, the attacker creates a "fork" of the blockchain, starting by mining another version of block 270,000 pointing to the same block 269,999 as a parent but with the new transaction in place of the old one. Because the block data is different, this requires redoing the proof of work. Furthermore, the attacker's new version of block 270,000 has a different hash, so the original blocks 270,001 to 270,005 do not "point" to it; thus, the original chain and the attacker's new chain are completely separate. The rule is that in a fork the longest blockchain is taken to be the truth, and so legitimate miners will work on the 270,005 chain while the attacker alone is working on the 270,000 chain. In order for the attacker to make his blockchain the longest, he would need to have more computational power than the rest of the network combined in order to catch up (hence, "51% attack").

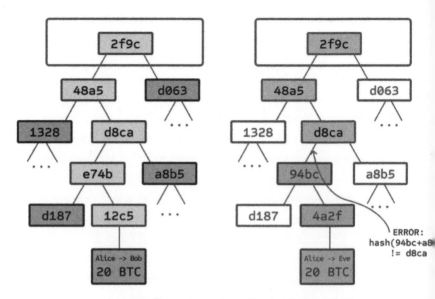

Left: it suffices to present only a small number of nodes in a Merkle tree to give a proof of the validity of a branch.
Right: any attempt to change any part of the Merkle tree will eventually lead to an inconsistency somewhere up the chain.

MERKLE TREES

An important scalability feature of Bitcoin is that the block is stored in a multi-level data structure. The "hash" of a block is actually only the hash of the block header, a roughly two-hundred-byte piece of data that contains the timestamp, nonce, previous block hash, and the root hash of a data structure called the Merkle tree storing all transactions in the block. A Merkle tree is a type of binary tree, composed of a set of nodes with a large number of leaf nodes at the bottom of the tree containing the underlying data, a set of intermediate nodes where each node is the hash of its two children, and finally a single root node, also formed from the hash of its two children, representing the "top" of the tree.

The purpose of the Merkle tree is to allow the data in a block to be delivered piecemeal: a node can download only the header of a block from one source, the small part of the tree relevant to them from another source, and still be assured that all of the data is correct. The reason why this works is that hashes propagate upward: if a malicious user attempts to swap in a fake transaction into the bottom of a Merkle tree, this change will cause a change in the node above, and then a change in the node above that, finally changing the root of the tree and therefore the hash of the block, causing the protocol to register it as a completely different block (almost certainly with an invalid proof of work).

The Merkle tree protocol is arguably essential to long-term sustainability. A "full node" in the Bitcoin network, one that stores and processes the entirety of every block, takes up about 15 GB of disk space in the Bitcoin network as of April 2014, and is growing by over a gigabyte per month. Currently, this is viable for some desktop computers and not phones, and later on in the future only businesses and hobbyists will be able to participate. A protocol known as "simplified payment verification" (SPV) allows for another class of nodes to exist, called "light nodes," which download the block headers, verify the proof of work on the block headers, and then download only the "branches" associated with transactions that are relevant to them. This allows light nodes to determine with a strong guarantee of security what the status of any Bitcoin transaction, and their current balance, is while downloading only a very small portion of the entire blockchain.

ALTERNATIVE BLOCKCHAIN APPLICATIONS

The idea of taking the underlying blockchain idea and applying it to other concepts also has a long history. In 2005, Nick Szabo came out with the concept of "secure property titles with owner authority," a document describing how "new advances in replicated database technology" will allow for a blockchain-based system for storing a registry of who owns what land, creating an elaborate framework including concepts such as homesteading, adverse possession, and Georgian land tax. However, there was unfortunately no effective replicated database system available at the time, and so the protocol was never implemented in practice. After 2009, however, once Bitcoin's decentralized consensus was developed, a number of alternative applications rapidly began to emerge.

◻ **NAMECOIN:** Created in 2010, Namecoin is best described as a decentralized name-registration database. In decentralized protocols like Tor, Bitcoin, and BitMessage, there needs to be some way of identifying accounts so that other people can interact with them, but in all existing solutions the only kind of identifier available is a pseudo-random hash like `1LW79wp5ZBqaHW1jL5TCiBCrhQYtHagU-Wy`. Ideally, one would like to be able to have an account with a name like "george." However, the problem is that if one person can create an account named "george" then someone else can use the same process to register "george" for themselves as well and impersonate them. The only solution is a first-to-file paradigm, where the first registerer succeeds and the second fails—a problem perfectly suited for the Bitcoin consensus protocol. Namecoin is the oldest, and most successful, implementation of a name registration system using such an idea.

◻ **COLORED COINS:** The purpose of colored coins is to serve as a protocol to allow people to create their own digital currencies—or, in the important trivial case of a currency with one unit, digital tokens—on the Bitcoin blockchain. In the colored coins protocol, one "issues" a new currency by publicly assigning a color to a specific Bitcoin UTXO, and the protocol recursively defines the color of other UTXO to be the same as the color of the inputs that the transaction creating them spent (some special rules apply in the case of mixed-color inputs). This allows users to maintain wallets containing only UTXO of a specific color and send them around much like regular bitcoins, backtracking through the blockchain to determine the color of any UTXO that they receive.

□ **METACOINS:** The idea behind a metacoin is to have a protocol that lives on top of Bitcoin, using Bitcoin transactions to store metacoin transactions but having a different state transition function, APPLY'. Because the metacoin protocol cannot prevent invalid metacoin transactions from appearing in the Bitcoin blockchain, a rule is added that if APPLY'(S,TX) returns an error, the protocol defaults to APPLY'(S,TX) = S. This provides an easy mechanism for creating an arbitrary cryptocurrency protocol, potentially with advanced features that cannot be implemented inside Bitcoin itself, but with a very low development cost since the complexities of mining and networking are already handled by the Bitcoin protocol. Metacoins have been used to implement some classes of financial contracts, name registration, and decentralized exchange.

Thus, in general, there are two approaches toward building a consensus protocol: building an independent network, and building a protocol on top of Bitcoin. The former approach, while reasonably successful in the case of applications like Namecoin, is difficult to implement; each individual implementation needs to bootstrap an independent blockchain, as well as building and testing all of the necessary state transition and networking code. Additionally, we predict that the set of applications for decentralized consensus technology will follow a power law distribution where the vast majority of applications would be too small to warrant their own blockchain, and we note that there exist large classes of decentralized applications, particularly decentralized autonomous organizations, that need to interact with each other.

The Bitcoin-based approach, on the other hand, has the flaw that it does not inherit the simplified payment-verification features of Bitcoin. SPV works for Bitcoin because it can use blockchain depth as a proxy for validity; at some point, once the ancestors of a transaction go far enough back, it is safe to say that they were legitimately part of the state. Blockchain-based meta-protocols, on the other hand, cannot force the blockchain not to include transactions that are not valid within the context of their own protocols. Hence, a fully secure SPV meta-protocol implementation would need to backward scan all the way to the beginning of the Bitcoin blockchain to determine whether or not certain transactions are valid. Currently, all "light" implementations of Bitcoin-based meta-protocols rely on a trusted server to provide the data, arguably a highly suboptimal result especially when one of the primary purposes of a cryptocurrency is to eliminate the need for trust.

SCRIPTING

Even without any extensions, the Bitcoin protocol actually does facilitate a weak version of a concept of "smart contracts." UTXO in Bitcoin can be owned not just by a public key, but also by a more complicated script expressed in a simple stack-based programming language. In this paradigm, a transaction spending that UTXO must provide data that satisfies the script. Indeed, even the basic public key ownership mechanism is implemented via a script: the script takes an elliptic curve signature as input, verifies it against the transaction and the address that owns the UTXO, and returns 1 if the verification is successful and 0 otherwise. Other, more complicated, scripts exist for various additional use cases. For example, one can construct a script that requires signatures from two out of a given three private keys to validate ("multisig"), a setup useful for corporate accounts, secure savings accounts, and some merchant escrow situations. Scripts can also be used to pay bounties for solutions to computational problems, and one can even construct a script that says something like "this Bitcoin UTXO is yours if you can provide an SPV proof that you sent a Dogecoin transaction of this denomination to me," essentially allowing decentralized cross-cryptocurrency exchange.

However, the scripting language as implemented in Bitcoin has several important limitations:

□ **LACK OF TURING-COMPLETENESS:** That is to say, while there is a large subset of computation that the Bitcoin scripting language supports, it does not nearly support everything. The main category that is missing is loops. This is done to avoid infinite loops during transaction verification; theoretically it is a surmountable obstacle for script programmers, since any loop can be simulated by simply repeating the underlying code many times with an if statement, but it does lead to scripts that are very space-inefficient. For example, implementing an alternative elliptic curve signature algorithm would likely require 256 repeated multiplication rounds all individually included in the code.

☐ **VALUE-BLINDNESS:** There is no way for a UTXO script to provide fine-grained control over the amount that can be withdrawn. For example, one powerful use case of an oracle contract would be a hedging contract, where A and B put in $1,000 worth of BTC and after thirty days the script sends $1,000 worth of BTC to A and the rest to B. This would require an oracle to determine the value of 1 BTC in USD, but even then it is a massive improvement in terms of trust and infrastructure requirement over the fully centralized solutions that are available now. However, because UTXO are all-or-nothing, the only way to achieve this is through the very inefficient hack of having many UTXO of varying denominations (e.g., one UTXO of $2k$ for every k up to 30) and having O pick which UTXO to send to A and which to B.

☐ **LACK OF STATE:** UTXO can either be spent or unspent; there is no opportunity for multi-stage contracts or scripts which keep any other internal state beyond that. This makes it hard to make multi-stage-options contracts, decentralized exchange offers, or two-stage cryptographic commitment protocols (necessary for secure computational bounties). It also means that UTXO can only be used to build simple, one-off contracts and not more complex "stateful" contracts such as decentralized organizations, and makes meta-protocols difficult to implement. Binary state combined with value-blindness also mean that another important application, withdrawal limits, is impossible.

☐ **BLOCKCHAIN-BLINDNESS:** UTXO are blind to blockchain data such as the nonce, the timestamp, and previous block hash. This severely limits applications in gambling, and several other categories, by depriving the scripting language of a potentially valuable source of randomness.

Thus, we see three approaches to building advanced applications on top of cryptocurrency: building a new blockchain, using scripting on top of Bitcoin, and building a meta-protocol on top of Bitcoin. Building a new blockchain allows for unlimited freedom in building a feature set, but at the cost of development time, bootstrapping effort, and security. Using scripting is easy to implement and standardize, but is very limited in its capabilities, and meta-protocols, while easy, suffer from faults in scalability. With Ethereum, we intend to build an alternative framework that provides even larger gains in ease of development as well as even stronger light client properties, while at the same time allowing applications to share an economic environment and blockchain security.

ETHEREUM

The intent of Ethereum is to create an alternative protocol for building decentralized applications, providing a different set of tradeoffs that we believe will be very useful for a large class of decentralized applications, with particular emphasis on situations where rapid development time, security for small and rarely used applications, and the ability of different applications to very efficiently interact, are important. Ethereum does this by building what is essentially the ultimate abstract foundational layer: a blockchain with a built-in Turing-complete programming language, allowing anyone to write smart contracts and decentralized applications where they can create their own arbitrary rules for ownership, transaction formats, and state transition functions. A bare-bones version of Namecoin can be written in two lines of code, and other protocols like currencies and reputation systems can be built in under twenty. Smart contracts, cryptographic "boxes" that contain value and only unlock it if certain conditions are met, can also be built on top of the platform, with vastly more power than that offered by Bitcoin scripting because of the added powers of Turing-completeness, value-awareness, blockchain-awareness, and state.

ETHEREUM ACCOUNTS

In Ethereum, the state is made up of objects called "accounts," with each account having a twenty-byte address and state transitions being direct transfers of value and information between accounts. An Ethereum account contains four fields:

☐ The **nonce**, a counter used to make sure each transaction can only be processed once

☐ The account's current **ether balance**

☐ The account's **contract code**, if present

☐ The account's **storage** (empty by default)

"Ether" is the main internal crypto-fuel of Ethereum, and is used to pay transaction fees. In general, there are two types of accounts: **externally owned accounts**, controlled by private keys, and **contract accounts**, controlled by their contract code. An externally owned account has no code, and one can send messages from an externally owned account by creating and signing a transaction; in a contract account, every time the contract account receives a message its code activates, allowing it to read and write to internal storage and send other messages or create contracts in turn.

Note that "contracts" in Ethereum should not be seen as something that should be "fulfilled" or "complied with"; rather, they are more like "autonomous agents" that live inside of the Ethereum execution environment, always executing a specific piece of code when "poked" by a message or transaction, and having direct control over their own ether balance and their own key/value store to keep track of persistent variables.

MESSAGES AND TRANSACTIONS

The term "transaction" is used in Ethereum to refer to the signed data package that stores a message to be sent from an externally owned account. Transactions contain:

- □ The recipient of the message

- □ A signature identifying the sender

- □ The amount of ether to transfer from the sender to the recipient

- □ An optional data field

- □ A STARTGAS value, representing the maximum number of computational steps the transaction execution is allowed to take

- □ A GASPRICE value, representing the fee the sender pays per computational step

The first three are standard fields expected in any cryptocurrency. The data field has no function by default, but the virtual machine has an opcode using which a contract can access the data; as an example use case, if a contract is functioning as an on-blockchain domain-registration service, then it may wish to interpret the data being passed to it as containing two "fields," the first field being a domain to register and the second field being the IP address to register it to. The contract would read these values from the message data and appropriately place them in storage.

The STARTGAS and GASPRICE fields are crucial for Ethereum's anti-denial-of-service model. In order to prevent accidental or hostile infinite loops or other computational wastage in code, each transaction is required to set a limit to how many computational steps of code execution it can use. The fundamental unit of computation is "gas"; usually, a computational step costs 1 gas, but some operations cost higher amounts of gas because they are more computationally expensive, or increase the amount of data that must be stored as part of the state. There is also a fee of 5 gas for every byte in the transaction data. The intent of the fee system is to require an attacker to pay proportionately for every resource that they consume, including computation, bandwidth, and storage; hence, any transaction that leads to the network consuming a greater amount of any of these resources must have a gas fee roughly proportional to the increment.

MESSAGES

Contracts have the ability to send "messages" to other contracts. Messages are virtual objects that are never serialized and exist only in the Ethereum execution environment. A message contains:

- □ The sender of the message (implicit)
- □ The recipient of the message
- □ The amount of ether to transfer alongside the message
- □ An optional data field

A **STARTGAS** VALUE

Essentially, a message is like a transaction, except it is produced by a contract and not an external actor. A message is produced when a contract currently executing code executes the CALL opcode, which produces and executes a message. Like a transaction, a message leads to the recipient account running its code. Thus, contracts can have relationships with other contracts in exactly the same way that external actors can.

Note that the gas allowance assigned by a transaction or contract applies to the total gas consumed by that transaction and all sub-executions. For example, if an external actor A sends a transaction to B with 1,000 gas, and B consumes 600 gas before sending a message to C, and the internal execution of C consumes 300 gas before returning, then B can spend another 100 gas before running out of gas.

ETHEREUM STATE TRANSITION FUNCTION

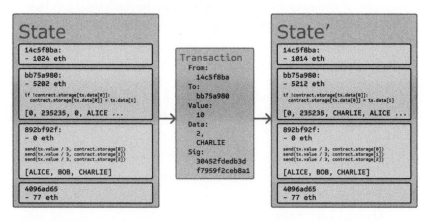

The Ethereum state transition function, APPLY(S, TX) -> S' can be defined as follows:

1. Check if the transaction is well-formed (i.e., has the right number of values), the signature is valid, and the nonce matches the nonce in the sender's account. If not, return an error.

2. Calculate the transaction fee as STARTGAS * GASPRICE, and determine the sending address from the signature. Subtract the fee from the sender's account balance and increment the sender's nonce. If there is not enough balance to spend, return an error.

3. Initialize GAS = STARTGAS, and take off a certain quantity of gas per byte to pay for the bytes in the transaction.

4. Transfer the transaction value from the sender's account to the receiving account. If the receiving account does not yet exist, create it. If the receiving account is a contract, run the contract's code either to completion or until the execution runs out of gas.

5. If the value transfer failed because the sender did not have enough money, or the code execution ran out of gas, revert all state changes except the payment of the fees, and add the fees to the miner's account.

6. Otherwise, refund the fees for all remaining gas to the sender, and send the fees paid for gas consumed to the miner.

For example, suppose that the contract's code is:

```
if !self.storage[calldataload(0)]:
    self.storage[calldataload(0)] = calldataload(32)
```

Note that in reality the contract code is written in the low-level EVM code; this example is written in Serpent, one of our high-level languages, for clarity, and can be compiled down to EVM code. Suppose that the contract's storage starts off empty, and a transaction is sent with 10 ether value, 2,000 gas, 0.001 ether gasprice, and 64 bytes of data, with bytes 0–31 representing the number 2 and bytes 32–63 representing the string CHARLIE.* The process for the state transition function in this case is as follows:

1. Check that the transaction is valid and well formed.

2. Check that the transaction sender has at least 2,000 × 0.001 = 2 ether. If it is, then subtract 2 ether from the sender's account.

3. Initialize gas = 2000; assuming the transaction is 170 bytes long and the byte-fee is 5, subtract 850 so that there is 1,150 gas left.

4. Subtract 10 more ether from the sender's account, and add it to the contract's account.

5. Run the code. In this case, this is simple: it checks if the contract's storage at index 2 is used, notices that it is not, and so it sets the storage at index 2 to the value CHARLIE. Suppose this takes 187 gas, so the remaining amount of gas is 1150 – 187 = 963.

6. Add 963 × 0.001 = 0.963 ether back to the sender's account, and return the resulting state.

If there was no contract at the receiving end of the transaction,

---

* *In original:* Internally, 2 and CHARLIE are both numbers, with the latter being in big-endian base 256 representation. Numbers can be at least 0 and at most 2256-1.

then the total transaction fee would simply be equal to the provided GASPRICE multiplied by the length of the transaction in bytes, and the data sent alongside the transaction would be irrelevant.

Note that messages work equivalently to transactions in terms of reverts: if a message execution runs out of gas, then that message's execution, and all other executions triggered by that execution, revert, but parent executions do not need to revert. This means that it is "safe" for a contract to call another contract, as if A calls B with g gas then A's execution is guaranteed to lose at most g gas. Finally, note that there is an opcode, CREATE, that creates a contract; its execution mechanics are generally similar to CALL, with the exception that the output of the execution determines the code of a newly created contract.

CODE EXECUTION

The code in Ethereum contracts is written in a low-level, stack-based bytecode language, referred to as "Ethereum virtual machine code" or "EVM code." The code consists of a series of bytes, where each byte represents an operation. In general, code execution is an infinite loop that consists of repeatedly carrying out the operation at the current program counter (which begins at zero) and then incrementing the program counter by one, until the end of the code is reached or an error or STOP or RETURN instruction is detected. The operations have access to three types of space in which to store data:

◻ The **stack**, a last-in-first-out container to which values can be pushed and popped.

◻ **Memory**, an infinitely expandable byte array.

◻ The contract's long-term **storage**, a key/value store. Unlike stack and memory, which reset after computation ends, storage persists for the long term.

The code can also access the value, sender, and data of the incoming message, as well as block header data, and the code can also return a byte array of data as an output.

The formal execution model of EVM code is surprisingly simple. While the Ethereum virtual machine is running, its full computational state can be defined by the tuple (`block_state`, `transaction`, `message`, `code`, `memory`, `stack`, `pc`, `gas`), where `block_state` is the global state containing all accounts and includes balances and storage. At the start of every round of execution, the current instruction is found by taking the pcth byte of `code` (or 0 if `pc >= len(code)`), and each instruction has its own definition in terms of how it affects the tuple. For example, `ADD` pops two items off the stack and pushes their sum, reduces `gas` by 1 and increments `pc` by 1, and `SSTORE` pushes the top two items off the stack and inserts the second item into the contract's storage at the index specified by the first item. Although there are many ways to optimize Ethereum virtual-machine execution via just-in-time compilation, a basic implementation of Ethereum can be done in a few hundred lines of code.

BLOCKCHAIN AND MINING

The Ethereum blockchain is in many ways similar to the Bitcoin blockchain, although it does have some differences. The main difference between Ethereum and Bitcoin with regard to the blockchain architecture is that, unlike Bitcoin, Ethereum blocks contain a copy of both the transaction list and the most recent state. Aside from that, two other values, the block number and the difficulty, are also stored in the block. The basic block validation algorithm in Ethereum is as follows:

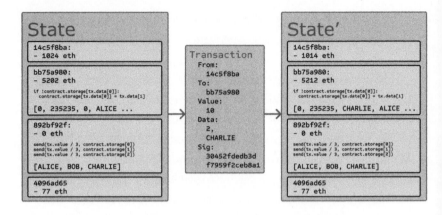

1. Check if the previous block referenced exists and is valid.

2. Check that the timestamp of the block is greater than that of the referenced previous block and less than fifteen minutes into the future.

3. Check that the block number, difficulty, transaction root, uncle root, and gas limit (various low-level Ethereum-specific concepts) are valid.

4. Check that the proof of work on the block is valid.

5. Let S[0] be the state at the end of the previous block.

6. Let TX be the block's transaction list, with n transactions. For all i in 0...n-1, set S[i+1] = APPLY(S[i],TX[i]). If any applications return an error, or if the total gas consumed in the block up until this point exceeds the GAS-LIMIT, return an error.

7. Let S_FINAL be S[n], but adding the block reward paid to the miner.

8. Check if the Merkle tree root of the state S_FINAL is equal

to the final state root provided in the block header. If it is, the block is valid; otherwise, it is not valid.

The approach may seem highly inefficient at first glance, because it needs to store the entire state with each block, but in reality efficiency should be comparable to that of Bitcoin. The reason is that the state is stored in the tree structure, and after every block only a small part of the tree needs to be changed. Thus, in general, between two adjacent blocks the vast majority of the tree should be the same, and therefore the data can be stored once and referenced twice using pointers (i.e., hashes of subtrees). A special kind of tree known as a "Patricia tree" is used to accomplish this, including a modification to the Merkle tree concept that allows for nodes to be inserted and deleted, and not just changed, efficiently. Additionally, because all of the state information is part of the last block, there is no need to store the entire blockchain history—a strategy which, if it could be applied to Bitcoin, can be calculated to provide 5–20x savings in space.

A commonly asked question is "where" contract code is executed, in terms of physical hardware. This has a simple answer: the process of executing contract code is part of the definition of the state transition function, which is part of the block-validation algorithm, so if a transaction is added into block B the code execution spawned by that transaction will be executed by all nodes, now and in the future, that download and validate block B.

APPLICATIONS

In general, there are three types of applications on top of Ethereum. The first category is financial applications, providing users with more powerful ways of managing and entering into contracts using their money. This includes sub-currencies, financial derivatives, hedging contracts, savings wallets, wills, and ultimately even some classes of full-scale employment contracts. The second category is semi-financial applications, where money is involved but there is also a heavy non-monetary side to what is being done; a perfect example is self-enforcing bounties for solutions to computational problems. Finally, there are applications such as online voting and decentralized governance that are not financial at all.

TOKEN SYSTEMS

On-blockchain token systems have many applications ranging from sub-currencies representing assets such as USD or gold to company stocks, individual tokens representing smart property, secure unforgeable coupons, and even token systems with no ties to conventional value at all, used as point systems for incentivization. Token systems are surprisingly easy to implement in Ethereum. The key point to understand is that all a currency, or token system, fundamentally is, is a database with one operation: subtract x units from A and give x units to B, with the proviso that (i) A had at *least* x units before the transaction and (2) the transaction is approved by A. All that it takes to implement a token system is to implement this logic into a contract.

The basic code for implementing a token system in Serpent looks as follows:

```
def send(to, value):
    if self.storage[msg.sender] >= value:
        self.storage[msg.sender] = self.storage[msg.
sender] - value
        self.storage[to] = self.storage[to] + value
```

This is essentially a literal implementation of the "banking system" state transition function described further above in this document. A few extra lines of code need to be added to provide for the initial step of distributing the currency units in the first place and a few other edge cases, and ideally a function would be added to let other contracts query for the balance of an address. But that's all there is to it. Theoretically, Ethereum-based token systems acting as sub-currencies can potentially include another important feature that on-chain Bitcoin-based meta-currencies lack: the ability to pay transaction fees directly in that currency. The way this would be implemented is that the contract would maintain an ether balance with which it would refund ether used to pay fees to the sender, and it would refill this balance by collecting the internal currency units that it takes in fees and reselling them in a constant running auction. Users would thus need to "activate" their accounts with ether, but once the ether is there it would be reusable because the contract would refund it each time.

FINANCIAL DERIVATIVES AND STABLE-VALUE CURRENCIES

Financial derivatives are the most common application of a "smart contract," and one of the simplest to implement in code. The main challenge in implementing financial contracts is that the majority of them require reference to an external price ticker; for example, a very desirable application is a smart contract that hedges against the volatility of ether (or another cryptocurrency) with respect to the US dollar, but doing this requires the contract to know what the value of ETH/USD is. The simplest way to do this is through a "data feed" contract maintained by a specific party (e.g., NASDAQ) designed so that that party has the ability to update the contract as needed, and providing an interface that allows other contracts to send a message to that contract and get back a response that provides the price.

Given that critical ingredient, the hedging contract would look as follows:

1. Wait for party A to input 1,000 ether.

2. Wait for party B to input 1,000 ether.

3. Record the USD value of 1,000 ether, calculated by querying the data feed contract, in storage, say this is x.

4. After thirty days, allow A or B to "reactivate" the contract in order to send x worth of ether (calculated by querying the data feed contract again to get the new price) to A and the rest to B.

Such a contract would have significant potential in crypto-commerce. One of the main problems cited about cryptocurrency is the fact that it's volatile; although many users and merchants may want the security and convenience of dealing with cryptographic assets, they may not wish to face that prospect of losing 23% of the value of their funds in a single day. Up until now, the most commonly proposed solution has been issuer-backed assets; the idea is that an issuer creates a sub-currency in which they have the right to issue and revoke units, and provide one unit of the currency to anyone who provides them (offline) with one unit of a specified underlying asset (e.g., gold, USD). The issuer then promises to provide one unit of the underlying asset to anyone who sends back one unit of the crypto asset. This mechanism allows any non-cryptographic asset to be "uplifted" into a cryptographic asset, provided that the issuer can be trusted.

In practice, however, issuers are not always trustworthy, and in some cases the banking infrastructure is too weak, or too hostile, for such services to exist. Financial derivatives provide an alternative. Here, instead of a single issuer providing the funds to back up an asset, a decentralized market of speculators, betting that the price of a cryptographic reference asset (e.g., ETH) will go up, plays that role. Unlike issuers, speculators have no option to default on their side of the bargain because the hedging contract holds their funds in escrow. Note that this approach is not fully decentralized, because a trusted source is still needed to provide the price ticker, although arguably even still this is a massive improvement in terms of reducing infrastructure requirements (unlike being an issuer, issuing a price feed requires no licenses and can likely be categorized as free speech) and reducing the potential for fraud.

IDENTITY AND REPUTATION SYSTEMS

The earliest alternative cryptocurrency of all, Namecoin, attempted to use a Bitcoin-like blockchain to provide a name-registration system, where users can register their names in a public database alongside other data. The major cited use case is for a DNS system, mapping domain names like "bitcoin.org" (or, in Namecoin's case, "bitcoin.bit") to an IP address. Other use cases include email authentication and potentially more advanced reputation systems. Here is the basic contract to provide a Namecoin-like name registration system on Ethereum:

```
def register(name, value):
    if !self.storage[name]:
        self.storage[name] = value
```

The contract is very simple; all it is, is a database inside the Ethereum network that can be added to, but not modified or removed from. Anyone can register a name with some value, and that registration then sticks forever. A more sophisticated name registration contract will also have a "function clause" allowing other contracts to query it, as well as a mechanism for the "owner" (i.e., the first registerer) of a name to change the data or transfer ownership. One can even add reputation and web-of-trust functionality on top.

DECENTRALIZED FILE STORAGE

Over the past few years, there have emerged a number of popular online file storage startups, the most prominent being Dropbox, seeking to allow users to upload a backup of their hard drive and have the service store the backup and allow the user to access it in exchange for a monthly fee. However, at this point the file storage market is at times relatively inefficient; a cursory look at various existing solutions shows that, particularly at the "uncanny valley" 20–200 GB level at which neither free quotas nor enterprise-level discounts kick in, monthly prices for mainstream file storage costs are such that you are paying for more than the cost of the entire hard drive in a single month. Ethereum contracts can allow for the development of a decentralized file-storage ecosystem, where individual users can earn small quantities of money by renting out their own hard drives and unused space can be used to further drive down the costs of file storage.

The key underpinning piece of such a device would be what we have termed the "decentralized Dropbox contract." This contract works as follows. First, one splits the desired data up into blocks, encrypting each block for privacy, and builds a Merkle tree out of it. One then makes a contract with the rule that, every n blocks, the contract would pick a random index in the Merkle tree (using the previous block hash, accessible from contract code, as a source of randomness), and give x ether to the first entity to supply a transaction with a simplified payment-verification-like proof of ownership of the block at that particular index in the tree. When a user wants to re-download their file, they can use a micropayment channel protocol (e.g., pay 1 szabo per 32 kilobytes) to recover the file; the most fee-efficient approach is for the payer not to publish the transaction until the end, instead replacing the transaction with a slightly more lucrative one with the same nonce after every 32 kilobytes.

An important feature of the protocol is that, although it may seem like one is trusting many random nodes not to decide to forget the file, one can reduce that risk down to near-zero by splitting the file into many pieces via secret sharing, and watching the contracts to see whether each piece is still in some node's possession. If a contract is still paying out money, that provides a cryptographic proof that someone out there is still storing the file.

DECENTRALIZED AUTONOMOUS ORGANIZATIONS

The general concept of a "decentralized autonomous organization" is that of a virtual entity that has a certain set of members or shareholders which, perhaps with a 67% majority, have the right to spend the entity's funds and modify its code. The members would collectively decide on how the organization should allocate its funds. Methods for allocating a DAO's funds could range from bounties, salaries to even more exotic mechanisms such as an internal currency to reward work.

This essentially replicates the legal trappings of a traditional company or nonprofit but using only cryptographic blockchain technology for enforcement. So far much of the talk around DAOs has been around the "capitalist" model of a "decentralized autonomous corporation" (DAC) with dividend-receiving shareholders and tradable shares; an alternative, perhaps described as a "decentralized autonomous community," would have all members have an equal share in the decision-making and require 67% of existing members to agree to add or remove a member. The requirement that one person can only have one membership would then need to be enforced collectively by the group.

A general outline for how to code a DAO is as follows. The simplest design is simply a piece of self-modifying code that changes if two-thirds of members agree on a change. Although code is theoretically immutable, one can easily get around this and have de facto mutability by having chunks of the code in separate contracts, and having the address of which contracts to call stored in the modifiable storage. In a simple implementation of such a DAO contract, there would be three transaction types, distinguished by the data provided in the transaction:

- □ [0,i,K,V] to register a proposal with index i to change the address at storage index K to value V

- □ [1,i] to register a vote in favor of proposal i

- □ [2,i] to finalize proposal i if enough votes have been made

The contract would then have clauses for each of these. It would maintain a record of all open-storage changes, along with a list of who voted for them. It would also have a list of all members. When any storage change gets to two-thirds of members voting for it, a finalizing transaction could execute the change. A more sophisticated skeleton would also have built-in voting ability for features like sending a transaction, adding members, and removing members, and may even provide for Liquid Democracy-style vote delegation (i.e., anyone can assign someone to vote for them, and assignment is transitive so if A assigns B and B assigns C, then C determines A's vote). This design would allow the DAO to grow organically as a decentralized community, allowing people to eventually delegate the task of filtering out who is a member to specialists, although unlike in the "current system" specialists can easily pop in and out of existence over time as individual community members change their alignments.

An alternative model is for a decentralized corporation, where any account can have zero or more shares, and two-thirds of the shares are required to make a decision. A complete skeleton would involve asset-management functionality, the ability to make an offer to buy or sell shares, and the ability to accept offers (preferably with an order-matching mechanism inside the contract). Delegation would also exist Liquid Democracy-style, generalizing the concept of a "board of directors."

FURTHER APPLICATIONS

1. SAVINGS WALLETS: Suppose that Alice wants to keep her funds safe, but is worried that she will lose, or someone will hack, her private key. She puts ether into a contract with Bob, a bank, as follows:

- □ Alice alone can withdraw a maximum of 1% of the funds per day.

- □ Bob alone can withdraw a maximum of 1% of the funds per day, but Alice has the ability to make a transaction with her key shutting off this ability.

- □ Alice and Bob together can withdraw anything.

Normally, 1% per day is enough for Alice, and if Alice wants to withdraw more she can contact Bob for help. If Alice's key gets hacked, she runs to Bob to move the funds to a new contract. If she loses her key, Bob will get the funds out eventually. If Bob turns out to be malicious, then she can turn off his ability to withdraw.

2. CROP INSURANCE: One can easily make a financial derivatives contract but using a data feed of the weather instead of any price index. If a farmer in Iowa purchases a derivative that pays out inversely based on the precipitation in Iowa, then if there is a drought, the farmer will automatically receive money and if there is enough rain the farmer will be happy because their crops would do well. This can be expanded to natural-disaster insurance generally.

3. A DECENTRALIZED DATA FEED: For financial contracts for difference, it may actually be possible to decentralize the data feed via a protocol called "SchellingCoin." SchellingCoin basically works as follows: n parties all put into the system the value of a given datum (e.g., the ETH/USD price), the values are sorted, and everyone between the twenty-fifth and seventy-fifth percentile gets one token as a reward. Everyone has the incentive to provide the answer that everyone else will provide, and the only value that a large number of players can realistically agree on is the obvious default: the truth. This creates a decentralized protocol that can theoretically provide any number of values, including the ETH/USD price, the temperature in Berlin, or even the result of a particular hard computation.

4. SMART MULTISIGNATURE ESCROW: Bitcoin allows multisignature-transaction contracts where, for example, three out of a given five keys can spend the funds. Ethereum allows for more granularity;

for example, four out of five can spend everything, three out of five can spend up to 10% per day, and two out of five can spend up to 0.5% per day. Additionally, Ethereum multisig is asynchronous—two parties can register their signatures on the blockchain at different times and the last signature will automatically send the transaction.

5. CLOUD COMPUTING: The EVM technology can also be used to create a verifiable computing environment, allowing users to ask others to carry out computations and then optionally ask for proofs that computations at certain randomly selected checkpoints were done correctly. This allows for the creation of a cloud-computing market where any user can participate with their desktop, laptop, or specialized server, and spot-checking, together with security deposits, can be used to ensure that the system is trustworthy (i.e., nodes cannot profitably cheat). Although such a system may not be suitable for all tasks; tasks that require a high level of inter-process communication, for example, cannot easily be done on a large cloud of nodes. Other tasks, however, are much easier to parallelize; projects like SETI@home, folding@home, and genetic algorithms can easily be implemented on top of such a platform.

6. PEER-TO-PEER GAMBLING: Any number of peer-to-peer gambling protocols, such as Frank Stajano and Richard Clayton's Cyber-Dice, can be implemented on the Ethereum blockchain. The simplest gambling protocol is actually simply a contract for difference on the next block hash, and more advanced protocols can be built up from there, creating gambling services with near-zero fees that have no ability to cheat.

7. PREDICTION MARKETS: Provided an oracle or SchellingCoin, prediction markets are also easy to implement, and prediction markets together with SchellingCoin may prove to be the first mainstream application of futarchy as a governance protocol for decentralized organizations.

8. ON-CHAIN DECENTRALIZED MARKETPLACES, using the identity and reputation system as a base.

MISCELLANEA AND CONCERNS

MODIFIED GHOST IMPLEMENTATION

The "Greedy Heaviest Observed Subtree" (GHOST) protocol is an innovation first introduced by Yonatan Sompolinsky and Aviv Zohar in December 2013. The motivation behind GHOST is that blockchains with fast confirmation times currently suffer from reduced security due to a high stale rate—because blocks take a certain time to propagate through the network, if miner A mines a block and then miner B happens to mine another block before miner A's block propagates to B, miner B's block will end up wasted and will not contribute to network security. Furthermore, there is a centralization issue: if miner A is a mining pool with 30% hashpower and B has 10% hashpower, A will have a risk of producing a stale block 70% of the time (since the other 30% of the time A produced the last block and so will get mining data immediately) whereas B will have a risk of producing a stale block 90% of the time. Thus, if the block interval is short enough for the stale rate to be high, A will be substantially more efficient simply by virtue of its size. With these two effects combined, blockchains which produce blocks quickly are very likely to lead to one mining pool having a large enough percentage of the network hashpower to have de facto control over the mining process.

As described by Sompolinsky and Zohar, GHOST solves the first issue of network security loss by including stale blocks in the calculation of which chain is the "longest"; that is to say, not just the parent and further ancestors of a block, but also the stale descendants of the block's ancestor (in Ethereum jargon, "uncles") are added to the calculation of which block has the largest total proof of work backing it. To solve the second issue of centralization bias, we go beyond the protocol described by Sompolinsky and Zohar, and also provide block rewards to stales: a stale block receives 87.5% of its base reward, and the nephew that includes the stale block receives the remaining 12.5%. Transaction fees, however, are not awarded to uncles.

Ethereum implements a simplified version of GHOST which only goes down seven levels. Specifically, it is defined as follows:

- ☐ A block must specify a parent, and it must specify 0 or more uncles

- ☐ An uncle included in block B must have the following properties:
  - It must be a direct child of the kth generation ancestor of B, where $2 <= k <= 7$.
  - It cannot be an ancestor of B
  - An uncle must be a valid block header, but does not need to be a previously verified or even valid block
  - An uncle must be different from all uncles included in previous blocks and all other uncles included in the same block (non-double-inclusion)

- ☐ For every uncle U in block B, the miner of B gets an additional 3.125% added to its coinbase reward and the miner of U gets 93.75% of a standard coinbase reward.

This limited version of GHOST, with uncles includable only up to seven generations, was used for two reasons. First, unlimited

GHOST would include too many complications into the calculation of which uncles for a given block are valid. Second, unlimited GHOST with compensation as used in Ethereum removes the incentive for a miner to mine on the main chain and not the chain of a public attacker.

FEES

Because every transaction published into the blockchain imposes on the network the cost of needing to download and verify it, there is a need for some regulatory mechanism, typically involving transaction fees, to prevent abuse. The default approach, used in Bitcoin, is to have purely voluntary fees, relying on miners to act as the gatekeepers and set dynamic minimums. This approach has been received very favorably in the Bitcoin community, particularly because it is "market-based," allowing supply and demand between miners and transaction senders determine the price. The problem with this line of reasoning is, however, that transaction processing is not a market; although it is intuitively attractive to construe transaction processing as a service that the miner is offering to the sender, in reality every transaction that a miner includes will need to be processed by every node in the network, so the vast majority of the cost of transaction processing is borne by third parties and not the miner that is making the decision of whether or not to include it. Hence, tragedy-of-the-commons problems are very likely to occur.

However, as it turns out this flaw in the market-based mechanism, when given a particular inaccurate simplifying assumption, magically cancels itself out. The argument is as follows. Suppose that:

1. A transaction leads to k operations, offering the reward kR

to any miner that includes it where R is set by the sender and k and R are (roughly) visible to the miner beforehand.

2. An operation has a processing cost of C to any node (i.e., all nodes have equal efficiency)

3. There are N mining nodes, each with exactly equal processing power (i.e., 1/N of total)

4. No non-mining full nodes exist.

A miner would be willing to process a transaction if the expected reward is greater than the cost. Thus, the expected reward is kR/N since the miner has a 1/N chance of processing the next block, and the processing cost for the miner is simply kC. Hence, miners will include transactions where kR/N > kC, or R > NC. Note that R is the per-operation fee provided by the sender, and is thus a lower bound on the benefit that the sender derives from the transaction, and NC is the cost to the entire network together of processing an operation. Hence, miners have the incentive to include only those transactions for which the total utilitarian benefit exceeds the cost.

However, there are several important deviations from those assumptions in reality:

1. The miner does pay a higher cost to process the transaction than the other verifying nodes, since the extra verification time delays block propagation and thus increases the chance the block will become a stale.

2. There do exist non-mining full nodes.

3. The mining power distribution may end up radically inegalitarian in practice.

4. Speculators, political enemies, and crazies whose utility function includes causing harm to the network do exist, and they can cleverly set up contracts where their cost is much lower than the cost paid by other verifying nodes.

(1) provides a tendency for the miner to include fewer transactions, and (2) increases NC; hence, these two effects at least partially cancel each other out. How? (3) and (4) are the major issue; to solve them we simply institute a floating cap: no block can have more operations than BLK_LIMIT_FACTOR times the long-term exponential moving average. Specifically:

```
blk.oplimit = floor((blk.parent.oplimit \* (EMAFACTOR
- 1) +
floor(parent.opcount \* BLK\_LIMIT\_FACTOR)) / EMA\_
FACTOR)
```

BLK_LIMIT_FACTOR and EMA_FACTOR are constants that will be set to 65536 and 1.5 for the time being, but will likely be changed after further analysis.

There is another factor disincentivizing large block sizes in Bitcoin: blocks that are large will take longer to propagate, and thus have a higher probability of becoming stales. In Ethereum, highly gas-consuming blocks can also take longer to propagate both because they are physically larger and because they take longer to process the transaction state transitions to validate. This delay disincentive is a significant consideration in Bitcoin, but less so in Ethereum because of the GHOST protocol; hence, relying on regulated block limits provides a more stable baseline.

COMPUTATION AND TURING-COMPLETENESS

An important note is that the Ethereum virtual machine is Turing-complete; this means that EVM code can encode any computation that can be conceivably carried out, including infinite loops. EVM code allows looping in two ways. First, there is a JUMP instruction that allows the program to jump back to a previous spot in the code, and a JUMPI instruction to do conditional jumping, allowing for statements like while x < 27: x = x * 2. Second, contracts can call other contracts, potentially allowing for looping through recursion. This naturally leads to a problem: Can malicious users essentially shut miners and full nodes down by forcing them to enter into an infinite loop? The issue arises because of a problem in computer science known as the halting problem: there is no way to tell, in the general case, whether or not a given program will ever halt.

As described in the state transition section, our solution works by requiring a transaction to set a maximum number of computational steps that it is allowed to take, and if execution takes longer computation is reverted but fees are still paid. Messages work in the same way. To show the motivation behind our solution, consider the following examples:

- □ An attacker creates a contract which runs an infinite loop, and then sends a transaction activating that loop to the miner. The miner will process the transaction, running the infinite loop, and wait for it to run out of gas. Even though the execution runs out of gas and stops halfway through, the transaction is still valid and the miner still claims the fee from the attacker for each computational step.

- □ An attacker creates a very long infinite loop with the

intent of forcing the miner to keep computing for such a long time that by the time computation finishes a few more blocks will have come out and it will not be possible for the miner to include the transaction to claim the fee. However, the attacker will be required to submit a value for STARTGAS limiting the number of computational steps that execution can take, so the miner will know ahead of time that the computation will take an excessively large number of steps.

□ An attacker sees a contract with code of some form like send(A, contract.storage[A]); contract.storage[A] = 0, and sends a transaction with just enough gas to run the first step but not the second (i.e., making a withdrawal but not letting the balance go down). The contract author does not need to worry about protecting against such attacks, because if execution stops halfway through the changes get reverted.

□ A financial contract works by taking the median of nine proprietary data feeds in order to minimize risk. An attacker takes over one of the data feeds, which is designed to be modifiable via the variable-address-call mechanism described in the section on DAOs, and converts it to run an infinite loop, thereby attempting to force any attempts to claim funds from the financial contract to run out of gas. However, the financial contract can set a gas limit on the message to prevent this problem.

The alternative to Turing-completeness is Turing-incompleteness, where JUMP and JUMPI do not exist and only one copy of each contract is allowed to exist in the call stack at any given time. With this system, the fee system described and the uncertainties around the effectiveness of our solution might not be necessary, as the cost of executing a contract would be bounded above by its size.

Additionally, Turing-incompleteness is not even that big a limitation; out of all the contract examples we have conceived internally, so far only one required a loop, and even that loop could be removed by making twenty-six repetitions of a one-line piece of code. Given the serious implications of Turing-completeness, and the limited benefit, why not simply have a Turing-incomplete language? In reality, however, Turing-incompleteness is far from a neat solution to the problem. To see why, consider the following contracts:

```
C0: call(C1); call(C1);
C1: call(C2); call(C2);
C2: call(C3); call(C3);
...
C49: call(C50); call(C50);
C50: (run one step of a program and record the change
in storage)
```

Now, send a transaction to A. Thus, in fifty-one transactions, we have a contract that takes up 2^{50} computational steps. Miners could try to detect such logic bombs ahead of time by maintaining a value alongside each contract specifying the maximum number of computational steps that it can take, and calculating this for contracts calling other contracts recursively, but that would require miners to forbid contracts that create other contracts (since the creation and execution of all twenty-six contracts above could easily be rolled into a single contract). Another problematic point is that the address field of a message is a variable, so in general it may not even be possible to tell which other contracts a given contract will call ahead of time. Hence, all in all, we have a surprising conclusion: Turing-completeness is surprisingly easy to manage, and the lack of Turing-completeness is equally surprisingly difficult to manage unless the exact same controls are in place—but in that case why not just let the protocol be Turing-complete?

CURRENCY AND ISSUANCE

The Ethereum network includes its own built-in currency, ether, which serves the dual purpose of providing a primary liquidity layer to allow for efficient exchange between various types of digital assets and, more importantly, of providing a mechanism for paying transaction fees. For convenience and to avoid future argument (see the current mBTC/uBTC/satoshi debate in Bitcoin), the denominations will be pre-labeled:

□ 1: wei

□ 10^{12}: szabo

□ 10^{15}: finney

□ 10^{18}: ether

This should be taken as an expanded version of the concept of "dollars" and "cents," or "BTC" and "satoshi." In the near future, we expect "ether" to be used for ordinary transactions, "finney" for microtransactions, and "szabo" and "wei" for technical discussions around fees and protocol implementation; the remaining denominations may become useful later and should not be included in clients at this point.

The issuance model will be as follows:

□ Ether will be released in a currency sale at the price of 1,000–2,000 ether per BTC, a mechanism intended to fund the Ethereum organization and pay for development that has been used with success by other platforms such as Mastercoin and NXT. Earlier buyers will benefit from larger discounts. The BTC received from the sale will be used entirely to pay salaries and bounties to developers and invested into various for-profit and nonprofit projects in the Ethereum and cryptocurrency ecosystem.

- ☐ $0.099x$ the total amount sold (60,102,216 ETH) will be allocated to the organization to compensate early contributors and pay ETH-denominated expenses before the genesis block.

- ☐ $0.099x$ the total amount sold will be maintained as a long-term reserve.

- ☐ $0.26x$ the total amount sold will be allocated to miners per year forever after that point.

Group	At launch	After 1 year	After 5 years
Currency Units	1.198X	1.458X	2.498X
Purchasers	83.5%	68.6%	40.0%
Reserve spent pre-sale	8.26%	6.79%	3.96%
Reserve spent post-sale	8.26%	6.79%	3.96%
Miners	0%	17.8%	52.0%

Long-Term Supply Growth Rate (percent)

The two main choices in the above model are (1) the existence and size of an endowment pool, and (2) the existence of a permanently growing linear supply, as opposed to a capped supply as in Bitcoin. The justification of the endowment pool is

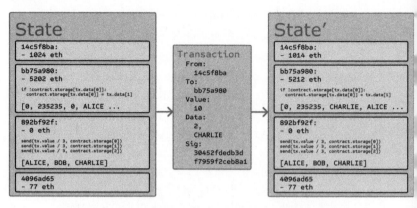

```
State
14c5f8ba:
- 1024 eth

bb75a980:
- 5202 eth
if !contract.storage[tx.data[0]]:
   contract.storage[tx.data[0]] = tx.data[1]

[0, 235235, 0, ALICE ...

892bf92f:
- 0 eth
send(tx.value / 3, contract.storage[0])
send(tx.value / 3, contract.storage[1])
send(tx.value / 3, contract.storage[2])

[ALICE, BOB, CHARLIE]

4096ad65
- 77 eth
```

```
Transaction
From:
   14c5f8ba
To:
   bb75a980
Value:
   10
Data:
   2,
   CHARLIE
Sig:
   30452fdedb3d
   f7959f2ceb8a1
```

```
State'
14c5f8ba:
- 1014 eth

bb75a980:
- 5212 eth
if !contract.storage[tx.data[0]]:
   contract.storage[tx.data[0]] = tx.data[1]

[0, 235235, CHARLIE, ALICE ...

892bf92f:
- 0 eth
send(tx.value / 3, contract.storage[0])
send(tx.value / 3, contract.storage[1])
send(tx.value / 3, contract.storage[2])

[ALICE, BOB, CHARLIE]

4096ad65
- 77 eth
```

Despite the linear currency issuance, just like with Bitcoin over time the supply growth
rate nevertheless tends to zero.

as follows. If the endowment pool did not exist, and the linear issuance reduced to $0.217x$ to provide the same inflation rate, then the total quantity of ether would be 16.5% less and so each unit would be 19.8% more valuable. Hence, in the equilibrium 19.8% more ether would be purchased in the sale, so each unit would once again be exactly as valuable as before. The organization would also then have $1.198x$ as much BTC, which can be considered to be split into two slices: the original BTC, and the additional $0.198x$. Hence, this situation is *exactly equivalent* to the endowment, but with one important difference: the organization holds purely BTC, and so is not incentivized to support the value of the ether unit.

The permanent linear supply-growth model reduces the risk of what some see as excessive wealth concentration in Bitcoin, and gives individuals living in present and future eras a fair chance to acquire currency units, while at the same time retaining a strong incentive to obtain and hold ether because the "supply-growth rate" as a percentage still tends to zero over time. We also theorize that because coins are always lost over time due to carelessness, death, etc., and coin loss can be modeled as a percentage of the total supply per year, that the total currency supply in circulation will in fact eventually stabilize at a value equal to the annual issuance divided by the loss rate (e.g., at a loss rate of 1%, once the supply reaches $26x$ then $0.26x$ will be mined and $0.26x$ lost every year, creating an equilibrium).

Note that in the future, it is likely that Ethereum will switch to a proof-of-stake model for security, reducing the issuance requirement to somewhere between zero and $0.05x$ per year. In the event that the Ethereum organization loses funding or for any other reason disappears, we leave open a "social contract": anyone has the right to create a future candidate version of Ethereum, with the only condition being that the quantity of ether must be at most equal to `60102216 * (1.198 + 0.26 * n)` where n is the number of years after the genesis block. Creators are free to crowd-sell or otherwise assign some or all of the difference between the PoS-driven supply expansion and the maximum allowable supply expansion to pay for development. Candidate upgrades that do not comply with the social contract may justifiably be forked into compliant versions.

MINING CENTRALIZATION

The Bitcoin mining algorithm works by having miners compute SHA256 on slightly modified versions of the block header millions of times over and over again, until eventually one node comes up with a version whose hash is less than the target (currently around 2^{192}). However, this mining algorithm is vulnerable to two forms of centralization. First, the mining ecosystem has come to be dominated by ASICs (application-specific integrated circuits), computer chips designed for, and therefore thousands of times more efficient at, the specific task of Bitcoin mining. This means that Bitcoin mining is no longer a highly decentralized and egalitarian pursuit, requiring millions of dollars of capital to effectively participate in. Second, most Bitcoin miners do not actually perform block validation locally; instead, they rely on a centralized mining pool to provide the block headers. This problem is arguably worse: as of the time of this writing, the top three mining pools indirectly control roughly 50% of processing power in the Bitcoin network, although this is mitigated by the fact that miners can switch to other mining pools if a pool or coalition attempts a 51% attack.

The current intent at Ethereum is to use a mining algorithm where miners are required to fetch random data from the state, compute some randomly selected transactions from the last N blocks in the blockchain, and return the hash of the result. This has two important benefits. First, Ethereum contracts can include any kind of computation, so an Ethereum ASIC would essentially be an ASIC for general computation—i.e., a better CPU. Second, mining requires access to the entire blockchain, forcing miners to store the entire blockchain and at least be capable of verifying every transaction. This removes the need for centralized mining pools; although mining pools can still serve the legitimate role of evening out the randomness of reward distribution, this function can be served equally well by peer-to-peer pools with no central control.

This model is untested, and there may be difficulties along the way in avoiding certain clever optimizations when using contract execution as a mining algorithm. However, one notably interesting feature of this algorithm is that it allows anyone to "poison the well," by introducing a large number of contracts into the blockchain specifically designed to stymie certain ASICs. The economic incentives exist for ASIC manufacturers to use such a trick to attack each other. Thus, the solution that we are developing is ultimately an adaptive economic human solution rather than purely a technical one.

SCALABILITY

One common concern about Ethereum is the issue of scalability. Like Bitcoin, Ethereum suffers from the flaw that every transaction needs to be processed by every node in the network. With Bitcoin, the size of the current blockchain rests at about 15 GB, growing by about 1 MB per hour. If the Bitcoin network were to process Visa's 2,000 transactions per second, it would grow by 1 MB per three seconds (1 GB per hour, 8 TB per year). Ethereum is likely to suffer a similar growth pattern, worsened by the fact that there will be many applications on top of the Ethereum blockchain instead of just a currency as is the case with Bitcoin, but ameliorated by the fact that Ethereum full nodes need to store just the state instead of the entire blockchain history.

The problem with such a large blockchain size is centralization risk. If the blockchain size increases to, say, 100 TB, then the likely scenario would be that only a very small number of large businesses would run full nodes, with all regular users using light SPV nodes. In such a situation, there arises the potential concern that the full nodes could band together and all agree to cheat in some profitable fashion (e.g., change the block reward, give themselves BTC). Light nodes would have no way of detecting this immediately. Of course, at least one honest full node would likely exist, and after a few hours information about the fraud would trickle out through channels like Reddit, but at that point it would be too late: it would be up to the ordinary users to organize an effort to blacklist the given blocks, a massive and likely infeasible coordination problem on a similar scale as that of pulling off a successful 51% attack. In the case of Bitcoin, this is currently a problem, but there exists a blockchain modification suggested by Peter Todd which will alleviate this issue.

In the near term, Ethereum will use two additional strategies to cope with this problem. First, because of the blockchain-based mining algorithms, at least every miner will be forced to be a full node, creating a lower bound on the number of full nodes. Second and more importantly, however, we will include an intermediate state tree root in the blockchain after processing each transaction. Even if block validation is centralized, as long as one honest verifying node exists, the centralization problem can be circumvented via a verification protocol. If a miner publishes an invalid block, that block must either be badly formatted, or the state S[n] is incorrect. Since S[0] is known to be correct, there must be some first state S[i] that is incorrect where S[i-1] is correct. The verifying node would provide the index i, along with a "proof of invalidity" consisting of the subset of Patricia tree nodes needing to process APPLY(S[i-1],TX[i]) -> S[i]. Nodes would be able to use those nodes to run that part of the computation, and see that the S[i] generated does not match the S[i] provided.

Another, more sophisticated, attack would involve the malicious miners publishing incomplete blocks, so the full information does not even exist to determine whether or not blocks are valid. The solution to this is a challenge-response protocol: verification nodes issue "challenges" in the form of target transaction indices, and upon receiving a node a light node treats the block as untrusted until another node, whether the miner or another verifier, provides a subset of Patricia nodes as a proof of validity.

CONCLUSION

The Ethereum protocol was originally conceived as an upgraded version of a cryptocurrency, providing advanced features such as on-blockchain escrow, withdrawal limits, financial contracts, gambling markets, and the like via a highly generalized programming language. The Ethereum protocol would not "support" any of the applications directly, but the existence of a Turing-complete programming language means that arbitrary contracts can theoretically be created for any transaction type or application. What is more interesting about Ethereum, however, is that the Ethereum protocol moves far beyond just currency. Protocols around decentralized file storage, decentralized computation, and decentralized prediction markets, among dozens of other such concepts, have the potential to substantially increase the efficiency of the computational industry, and provide a massive boost to other peer-to-peer protocols by adding for the first time an economic layer. Finally, there is also a substantial array of applications that have nothing to do with money at all.

The concept of an arbitrary state transition function as implemented by the Ethereum protocol provides for a platform with unique potential; rather than being a closed-ended, single-purpose protocol intended for a specific array of applications in data storage, gambling, or finance, Ethereum is open-ended by design, and we believe that it is extremely well-suited to serving as a foundational layer for a very large number of both financial and non-financial protocols in the years to come.

GLOSSARY

BLOCKCHAIN is the technology underlying Bitcoin, Ethereum, and similar **protocols**. A blockchain is a shared database whose contents the participating computers agree about. It is composed of blocks of data—containing transactions, software code, or other material—linked together as a continuous chain. Once added, data cannot be deleted or modified. The first blockchain is generally regarded to be that of Bitcoin, whose **genesis block** was mined on January 3, 2009.

CRYPTOCURRENCY is a general term for **blockchain**-based **tokens** that exhibit at least some (but usually not all) characteristics of traditional money, such as serving as a store of value or a medium of exchange. But rather than being backed by a government, cryptocurrencies generally gain adoption due to users' perceptions of their security, privacy, usability, or future market value.

CRYPTOECONOMICS is a paradigm frequently used in the design of **blockchain**-based systems, combining game theory, economic incentives, and **cryptographic** security. It is used to enable participants to coordinate around shared missions and products despite having little reason to trust one another.

CRYPTOGRAPHY is a field of mathematics and computer science, seeking to design secure communication and storage by encrypting data so it is accessible only to authorized users. Cryptographic techniques help make **blockchain** technology possible.

CYPHERPUNK is an ideology and political movement centered around using **cryptography** to increase personal privacy and freedom while reducing the power of governments to surveil and censor. Cypherpunk communities experimented for decades with ideas that became the basis of **blockchain** technology.

DAO stands for "decentralized autonomous organization," a term that generally refers to organizations that are to some degree defined by **smart contracts** on a **blockchain**. One of the first DAOs was "The DAO," an early Ethereum project whose June 2016 hack led to a "hard **fork**" of the Ethereum blockchain.

DAPP is short for "decentralized app"—any user-facing software that relies in some important way on interactions with **smart contracts** on a **blockchain**.

DECENTRALIZATION is a widely used concept in **blockchain** culture. While it has many possible meanings (see the chapter "The Meaning of Decentralization"), it generally refers to replacing systems under a single entity's control with systems that distribute control among their participants.

DEFI is short for "decentralized finance," or the phenomenon of creating financial instruments and software using **smart contracts** on **blockchains**. These include products for lending, earning interest, stable currencies, value transfer, and more.

ENS, or the Ethereum Name Service, is a registrar of unique domain names on the Ethereum **blockchain**, which can refer to wallet addresses. For instance, vitalik.eth is an ENS domain associated with one of the author's Ethereum addresses.

FORKING is the practice of copying open-source software code or data in order to modify it. This can be done for the purpose of releasing a parallel version or to improve an existing one. For instance, many early "altcoin" **cryptocurrencies** are forks of Bitcoin's software. Forking also refers to updates in the software for a **blockchain**, or when a single blockchain splits in two if some users adopt an update and others do not.

FUTARCHY is a system of governance, proposed by economist Robin Hanson, in which **prediction markets** determine the most effective policies for achieving commonly agreed-on goals.

GENESIS BLOCK refers to the first block of a **blockchain**. The term was first used in the context of Bitcoin and has been used since for Ethereum and other blockchains.

LAYERS 1 AND 2 in the context of **blockchains** refer to two types of network infrastructure. Layer 1 is the underlying blockchain **protocol**, such as Ethereum. Layer 2 includes intermediary services, such as **rollups**, that make running applications on the blockchain easier and cheaper.

MINING, in the context of **blockchain** systems that use **proof of work**, is the practice of using computational power to confirm new blocks of data and receive **token** rewards in return. While mining can be done by individual users, on many networks it is dominated by industrial operations involving large numbers of specialized computers and consuming considerable electricity.

NFT, or non-fungible token, refers to a class of **blockchain**-based **tokens** intended to be one-of-a-kind, as opposed to **cryptocurrencies**, in which all tokens are interchangeable. NFTs are often used to demonstrate ownership of artworks, digital assets, and community membership.

ON-CHAIN refers to activities that occur through direct interaction with a **blockchain**, such as a voting process using a **smart contract**. In contrast, off-chain activity might include deliberation about the vote on social media or a corporate board meeting to decide how to vote with the company's tokens.

ORACLES are systems that allow **smart contracts** to interact with the world outside their **blockchain**. For instance, an oracle might confirm that a certain news event occurred, or that a certain transaction on another blockchain was completed.

PEER-TO-PEER refers to a type of network made up of nodes that connect to each other as equals. Pre-**blockchain** examples include Napster and BitTorrent. This is in contrast to the client-server structure used for most websites and centralized platforms, where the server holds privileges that client users lack. Public blockchains like the Ethereum network allow any user to function as both client and server. Other kinds of blockchains, known as "permissioned," allow only certain users to act as peers.

PREDICTION MARKETS are systems that allow participants to bet on the outcomes of real-world events and be rewarded for bets that prove accurate. They are frequently more accurate than other forms of crowdsourcing and prediction.

PROOF OF STAKE is a method for appending data to a **blockchain** that requires **validator** computers on the network to "stake" **tokens** in order to participate in agreeing on which new data to accept and in what order. Validators receive token rewards for participating. The risk of losing staked tokens dissuades would-be attackers from attempting to corrupt the data.

PROOF OF WORK is a method for appending data to a **blockchain** that requires computers to carry out complex cryptographic calculations.

Greater processing power increases the chance of receiving rewards for **mining** a block. The cost of energy required to mine dissuades would-be attackers from attempting to corrupt the data.

PROTOCOLS are sets of rules for how computers interact with each other on a shared network. Protocols enable the internet (TCP/IP) and the web (HTTP); and **blockchain** networks such as Bitcoin and Ethereum are defined by protocols as well.

PUBLIC AND PRIVATE KEYS are strings of characters that form the basis of **cryptographic** systems. Any given address (similar to an account) on a **blockchain** can be accessed only with both the public key (similar to a username) and the private key (similar to a password).

PUBLIC GOODS is a concept in economics referring to things that can be used by anybody and whose use by one person does not exclude access for others. Examples include language, street lights, air, and open-source software. In the context of blockchain culture, public goods generally refer to software infrastructure that many parties rely on but that no one party owns or has sufficient incentive to develop.

QUADRATIC VOTING is a decision-making technique in which a user can allocate more **tokens** to influence a vote based on their wealth or the intensity of their preference. However, each additional token for a given user becomes more costly, in order to reduce the ability of a minority to easily overwhelm the majority. Proper functioning requires a robust system for confirming user identity.

ROLLUPS are intermediary systems that sit between users and an underlying **blockchain** as part of a **layer 2** ecosystem. They may offer features such as faster transactions and lower costs than the **layer 1** blockchain, while inheriting the security of layer 1. Rollups have become an important strategy for enabling Ethereum to scale beyond the capacity of its original design.

SCHELLING POINT, or focal point, refers to a conclusion that agents will tend to converge on when they cannot communicate with each other, often based on their predictions of how one another will act. Since a Schelling point often corresponds to the truth, the concept has been frequently used in the design of **oracles** and **prediction markets** in the context of **blockchains**. Its namesake is the Cold War game theorist Thomas Schelling.

SMART CONTRACTS are pieces of software designed to run on computational **blockchains** like Ethereum. A contract might carry out tasks like issuing **tokens**, enabling complex transactions, and prescribing a governance system.

TOKENS are units of value that can be defined according to a given **protocol**, or a **smart contract** on a **blockchain**. Some tokens might behave like currency, like shares of stock, or like a deed of ownership—all depending on how they are designed.

VALIDATORS are users in a **network** using **proof of stake** who can receive **token** rewards for validating transactions and adding blocks to the **blockchain**. They are required to "stake" tokens on the network, which they can lose if they do not perform their role properly.

ZERO-KNOWLEDGE PROOFS are a type of **cryptographic** technique that enables users to prove that they have certain information without providing that information itself, thus protecting the user's privacy.